PHILOSOPHY OF THE SCIENCES

PHILOSOPHY OF THE SCIENCES

OR THE RELATIONS BETWEEN THE DEPARTMENTS OF KNOWLEDGE

BY

F. R. TENNANT

D. D., B. Sc., Hon. D. D. (Oxon.)

Tarner Lecturer of Trinity College, Cambridge, for the year 1931–1932

CAMBRIDGE

AT THE UNIVERSITY PRESS

1932

CAMBRIDGE
UNIVERSITY PRESS

University Printing House, Cambridge CB2 8BS, United Kingdom

Cambridge University Press is part of the University of Cambridge.

It furthers the University's mission by disseminating knowledge in the pursuit of education, learning and research at the highest international levels of excellence.

www.cambridge.org
Information on this title: www.cambridge.org/9781107453500

First published 1932
First paperback edition 2014

A catalogue record for this publication is available from the British Library

ISBN 978-1-107-45350-0 Paperback

PREFACE

This Volume contains the Tarner Lectures for 1931–1932, substantially in the form in which they were delivered

F. R. TENNANT

CONTENTS

LECTURE VI

The Relation of Theology to other Departments of Knowledge

Lecture I

PHILOSOPHY OF THE SCIENCES

THE FIELD within which a subject may be chosen for a course of lectures on the Tarner foundation has been defined as "the philosophy of the sciences and the relations or want of relations between the different departments of knowledge". Within this large field there are, doubtless, questions as to the relation of one particular science or sphere of thought to some other, any single one of which might admit of discussion in sufficient detail to exhaust the half-dozen hours that lie at a lecturer's disposal. But rather than to deal with any such special problem I propose to take over the founder's formula, which I have quoted, in its comprehensiveness, as the title of the course which I have been invited to deliver. This will involve treatment of a wide subject with the brevity that courts the danger of superficiality or else of unintelligibility. I therefore hasten to add that I shall confine myself as far as possible to tracing but one among the several threads of connexion which might be followed, and that I shall attempt thus to relate only the broader divisions, and not the subdivisions, into which knowledge falls. The relation of our various knowledges —to revive an obsolete expression—in respect of their origination, their developement out of crude experience, and their exemplification, severally, of what knowledge consists in, may perhaps be profitably discussed without considering their respective contributions in facts and generalisations; and broad outlines may be sketched without referring to structural details and secondary interdependences. It is my hope that some service may thus be rendered at least to students at present devoting themselves more or less exclusively to some one of the many departments of knowledge by inviting their attention, during these spare hours, to certain relations in which typical sciences, or groups of sciences, stand to one another and to knowledge as a co-ordinated whole.

A discussion of this kind may at least serve to disclose continuity between diverse fields of thought where disparateness of method and independence as to deliverances had been presumed to subsist. Or, on the other hand, it may suggest that gaps, which had appeared to one's imagination as promising to be bridgeable by further advances in knowledge, are impassable, and that linkages which had seemed plausible are suspect or false. Thus, prepossessions of various kinds, which are naturally entertained so long as we adopt the standpoint and use the method of some particular sphere of knowledge, may become revised or discarded when we pass from the pursuit of a science to a survey, and—what is quite a different thing—a philosophy, of the sciences. What I mean by 'a philosophy of the sciences' will presently be defined; but, however variously the functions and scope of philosophy may be conceived, we shall all probably agree that one tendency of the study of it should be to prevent the confusion of standpoints and the collision of methods belonging to sciences differing as greatly as do, for instance, physiology, psychology, and theology. As a physiologist or a physicist one can regard the human being as if he were a complicated machine, the changes within which admit of analysis and even, theoretically speaking, of description in terms of equations; as a psychologist or a historian one can regard a man as a personal being, the essence of whose individuality has disappeared in the physicist's account of him; as a theologian one looks upon a man as a member of a supersensible world, the child and the image of God, and as affording the one clue to the end which the physical universe subserves. But it is only as a philosopher of the sciences that one can attempt to determine whether these several descriptions and estimations of the human being are compatible or incompatible with one another: independent or, in the last resort, mutually implicative. In this latter capacity alone can one fully discern that the sciences to which I have alluded are all merely witnesses to be cross-examined, and whose respective evidences are to be evaluated and co-ordinated in a court over which no special science, however great its prestige, can preside as judge.

This illustration will suffice to shew the need of a comprehensive survey of several departments of knowledge, and, indeed, of a further inquiry as to what 'knowing' precisely means within the sphere of each of them, if we would philosophise about life or think liberally upon the questions that are of most vital interest for humanity. Each of us has a personal need of a theory of the universe and of man's place in it. Such a theory, we may find, cannot be established by the strictest canons of scientific demonstration, but will necessarily combine elements of knowledge and faith, experience and hope, reason and feeling, certainty and probability. Its reasonableness will, therefore, depend upon the comprehensiveness of our *conspectus*. For the present, however, I will enlarge upon the converse truth that any one department of thought, in so far as it is not concerned with explicating its own presuppositions, ascertaining its limitations, or examining its relations in respect of interdependence, standpoint, data, method, etc. with other fields of thought, is precluded from supplying a foundation on which an intelligent belief as to man, the world, and God, can be securely based; and that, when any one department is so used, it is apt to lead to what may generically be called the 'specialist's fallacy'. Perhaps we shall later have occasion to observe the harmfulness to progress in sound knowledge which has resulted whenever philosophy has been for a time dominated, or has had its method prescribed to it, by any one of the special sciences.

With a view to gradually indicating the nature and value of a philosophy of the sciences I may begin by recalling some of the chief dangers attending what may be termed 'departmentality'. Firstly, in studying exclusively or predominantly some one branch of knowledge, whether it be historical or scientific or abstract, one may easily contract the tendency to treat the one aspect of the facts upon which that subject concentrates as if it were their only aspect. This, of course, is right and necessary in so far as knowledge of that subject alone is pursued, and then does but consist in ignoring what is irrelevant. And the same is true of the subject studied as well as of the student of it. For instance, science rightly eschews the invocation of final causes because such causes do not reveal Nature's *modus operandi*, with which science is solely concerned. Or, again, science is rightly disinterested, as we say, in the sense of being undistorted by the desire to impose human valuations on the world or its phenomena. It does not decide such a question of fact as the motions of heavenly bodies by declaring that planetary orbits must be circular because the circle is the perfect figure. Science, however, selects its data, and in doing so it omits all reference to those aspects of things in virtue of which they possess aesthetic and other kinds of value, or suggest purposefulness. And, these aspects having been ignored or eliminated in the selection of its data, it is not surprising that science should be silent about them when expounding its conclusions. But if the student of science, using his departmental knowledge as if it were a sufficient basis for a world-view, advances from the fact that physics abstracts from considerations as to value and purpose to the questionable beliefs that value is wholly subjective and that the world is meaningless, he does but assume that the aspect of the things he has studied is the only one that they possess. His philosophical conclusion should be that the only valuable philosophy is the one which avoids all questions concerning value. And his science, or rather science as he has allowed himself to misconceive it, would foreclose the gate to possible knowledge. It would preclude the investigation of matters to which it happens itself to be indifferent, and would then justly provoke the invitation to mind its own business. Thus it is apt to be hidden from the eye of the specialist how selective is the method used in his own subject, and how pertinent, when a wider field of view is to be surveyed, become aspects of fact which are irrelevant to that subject.

A second defect of departmentality in study and outlook is that it may incline the devotee of one field of knowledge to assume that there is disparity or discontinuity between certain things, and incompatibility or contradiction between certain deliverances or theories, whereas access to knowledge of other fields would mediate continuity or compatibility. And again, this applies to subjects of study, or sciences, if one may personify them for convenience. Perhaps an illustration or two will best shew what I now mean.

One instance of alleged discontinuity or disparateness that has been of outstanding significance and influence throughout the history of philosophy is that between sense and reason, and between the sciences which are concerned predominantly with the output of the one and the other of these faculties respectively. And closely connected with this seeming discontinuity is that between individual experience, with its particularity as to sensible quality, location, date, etc. and common or so-called 'universal' experience, such as scientific knowledge, which is shared by a plurality of persons, is independent of the idiosyncrasies of any individual's experience, and is characterised by generality, relative abstractness, and timelessness. It is but recently that the psychological side of the facts of mental heredity came to be investigated, and that a few philosophers have maintained that the comparing of notes by separate individuals and the consequent acquisition of a common standpoint, are of supreme importance for the study of the human knowledge-process. Before this light was forthcoming the discontinuities which I have mentioned naturally seemed to be real and incapable of being bridged. And so they do still to some who study analytic psychology in isolation from genetic psychology, who contemplate the finished products of reason and knowledge apart from the conditionings of common or shared experience, or who deny the relevance and importance of considerations concerning genesis and history in this connexion. But if I am not mistaken in my estimate of the significance of such considerations and their bearing on the validity of what purports to be knowledge, these influxes of psychological knowledge oblige us to renounce several venerable dogmas which have profoundly influenced, and, indeed, continue to influence, opinion as to the relations or want of relations between the different departments of knowledge. There will be opportunity later to discuss these issues; but for the present I will provisionally assert that there are reasons for believing that, so far from being disparate, sense and understanding have a common root; that sensory perception is not superfluous for the acquisition of higher knowledge or for cognitive access to ultimate reality; that faculties which used to be accounted innate in the individual subject, such as reason or conscience, are social products; that much that is wont to be regarded as immediate intuition is really mediate, and much that has passed for pure or *a priori* science is the outcome of empirically-inspired reasoning. Whether these statements, for the moment presented dogmatically, are capable of being substantiated or not is, however, not the most important matter in this context. I am just now only suggesting that knowledge such as the psychology of cognition *may* compel us to abandon an attitude, which another science has perhaps inclined us to adopt, towards a philosophical problem; and that our conception of knowledge in the abstract, and of concrete knowledge as a co-ordinated whole, must be affected by the inclusiveness or the departmentality of our intellectual pursuits. The piecemeal study of separate fields of thought may lead us to put questions wrongly and may supply wrong answers to questions rightly put.

Belief in discontinuities such as I have just referred to, and that perhaps do not exist, does not always arise directly from the limitations incidental to departmental learning. It may be the outcome of philosophical prejudice and personal temperament, inclining one to aversion from, or disparagement of, facts or sciences with which one is acquainted. This was the case with Kant, who was one of the greatest, if not the greatest, of philosophers, while the creature of his age and the slave of inbred preconceptions. Had Kant lived a century later he would probably have exercised his marvellous genius in seeking and establishing continuities with as much zeal as he actually drew hard lines of separation, and with more success than attended the indulgence of his craze for logical schematisms. Between sense and understanding, understanding and reason, practical reason and theoretical reason, impulse and moral freedom, belief and knowledge, Kant saw disparity or discontinuity; whereas we, partly owing to his hints as to possible connexions and partly in virtue of the rich suggestiveness of his very errors and his artificiality, have come rather to see actual links and stages of transition between the members of some at least of these pairs of processes and products.

Another instance of alleged disparateness which has influenced belief as to the unrelatableness of specific sciences is that between matter and mind, or, to use concepts of the same order and level, between matter and spirit. As they have come to be conceived for the purposes of common sense and science, matter and spirit seem, indeed, to have nothing in common. To matter we ascribe inertia and a causality from which the notion of efficiency is eliminable, while to spiritual beings we ascribe activity, that is to say, an actual *vis insita* as contrasted with a fictitious *vis a tergo*, or a subjective interest and striving which manifests itself in valuing, pursuing ends, and willing. Material changes are quantitative, theoretically reversible, and characterised by conservation; while in the mental domain we find quality irreducible to quantity, irreversibility, and no conservation-principles. Matter may conceivably be regarded as a *continuum* which we more or less arbitrarily differentiate into bodies or systems; but spirits or subjects cannot be conceived, compatibly with knowledge, save as an irreducible plurality; and so on. Belief in some of these differences, or rather in their ultimateness, and numerous problems thereby engendered, may, however, owe their existence to the departmentality of the studies of specialists, or to physics and psychology having each exclusively attended for long to its own data and business. In fact, this seems to some extent to have been the case. In what physicists commonly used to regard as wholly objective we can now discern subjective factors; and part of the data with which psychology is concerned, though once regarded by physicists and psychologists alike as purely subjective, are, in the most fundamental sense of the word, objective. Deeper insight into the problem set by observed facts than can be obtained from either physics or psychology, or from both, may be afforded by the science which, with unusual modesty, calls itself 'theory of knowledge'. This is the only science capable of conducting negotiations between the other two. And if this science is unable to proclaim the ultimate dualism of matter and spirit to be false, it is able to maintain that neither physics nor psychology would in any way be affected if matter were the appearance of humble spirit, and its inertia were the outcome of psychic activity; if, in other words, the discontinuity and disparity between matter and spirit were, from the point of view of metaphysics, non-existent. Thus once more we find that a verdict which one or more of the special sciences may be compelled to deliver, if they are to speak at all, on a question transcending the fields to which they have confined themselves, may need to be reversed when a linkage-science is called in for consultation. The philosophy which would see whole, we may now gather, is no mere sum of the sciences, each taken at its own valuation or without reservation and revision, but rather a co-ordination of such sciences in the condition in which they find themselves after being sifted by a critical theory or science of knowledge itself.

A propensity, the converse of that which I have been illustrating, may also beset departmentality in knowledge. I mean that just as apparent discontinuities may be acquiesced in through lack of acquaintance with such a science as may reveal unsuspected continuity, so also may identity or continuity be assumed or hoped for in cases where further knowledge might extinguish that hope or assumption. In our age the ideas of developement and universal continuity have proved so potent and fertile in directing thought and research as to become almost obsessive; wherefore it behoves us to be cautious lest we exaggerate the scope of their application. In this connexion I will briefly touch upon two gaps, as many will deem them, in the science of to-day, selected on account of their relevance to the subject of this course of lectures. The first is the discontinuity, real or apparent, between the inorganic and organic realms, and between the physical and the biological sciences. The presumable disparateness between these domains may possibly not suggest itself to a mind trained in the school of physical science. Such a mind may have become sanguine as to the all-sufficiency of the scientific or mechanistic method. It may feel assured that organic phenomena, though on the surface they differ from those described in works on physics or chemistry and refuse to illustrate the principle of the degradation of energy, will eventually admit of subsumption under physical laws without invocation of any but physical conceptions, vitalistic or psychological. This habit of mind is, of course, not universal among students of the physical sciences. It is, or at any rate was, more commonly possessed by biologists, who have less occasion than has the physicist to appreciate the abstractness and the limitations of the mechanistic method. But what I am concerned to remark, with regard to this tendency to see no gap where one presumably exists, is that it probably arises from inattention to questions which belong to the science of knowledge rather than to the domain of either physics or biology. The phrase 'mechanical explanation' is wont to be used in discussions of the reducibility of biology to physics as if it were unambiguous; and sometimes it is used as if it were synonymous with 'physico-chemical explanation'.

The phrase, however, admits of numerous interpretations. One of them is loose enough to meet the requirements of any branch of science, whatever special concepts it may need to use in addition to those supplied by mechanics; and one or two others are so rigid and pure as to be unemployable even in any department of physics. So long as this ambiguity is not resolved, and no preliminary inquiry is undertaken as to the difference between the most abstract kind of description and the various kinds of explanation, discussion as to the possibility of a mechanistic explanation of biological facts is inconclusive or, rather, futile. The question, whether living organisms have been evolved from lifeless matter, is not identical with the question, whether organisms are analysable into physico-chemical constituents; nor is the latter question the same as that as to whether organisms can be 'described' solely in terms of the concepts of chemistry and physics. But it is perhaps largely owing to these different issues being confounded that the bridging of the apparent gap between biology and mechanistic physics has seemed to some specialists to be only a matter of time, whereas to many philosophers it seems to involve the equating of what, on the hither side of metaphysics, are incommensurables.

The disparity between physical and mental events, as they are contemplated from the plane of science, constitutes the other great gap which prevents the reduction of all our knowledges to one unitary knowledge, such as that which presented itself to Mr Herbert Spencer in his incoherent dream of a synthetic philosophy composed of positive science. And the supposition that psychical occurrences can be evolved out of matter, as it is defined in mechanics, is nowadays not entertained, save for an exception here and there which perhaps one will in this place be allowed to neglect without reproof. But there is a narrower form of the obstacle presented by facts to the natural desire to establish continuity as far-reaching as can be desired, about which a few words will not be so plainly superfluous, and which I may select as another instance of the capacity of departmental knowledge to remain unaware of a gap such as comprehensive knowledge may detect. I refer to the familiar dictum that man is organic to Nature and an evolutionary product of the cosmos. This assertion embodies important truth; but there are reasons for regarding it as at the same time a half-truth. For it hushes up a discontinuity which, in the present state of our knowledge, cannot be resolved by any science. In a man's mental life, as a whole of connected experiences, there must be some connecting-thread in virtue of which those experiences are his, and his alone. The supposition that this thread is an abiding substance, or a soul, is perhaps the only one that explains the facts, and it is invoked on precisely the same grounds as is the continuous individuality which we ascribe to a tree, though the supposition of mental atoms so to speak, and of hypothetical laws as to their interactions, may suffice for a quasi-mathematical 'description' of the facts. But whether we prefer to call it a soul or not, the point is that this connecting-thread cannot be shewn to have been evolved, by variation and selection, from any humbler form of soul or thread. The continuity of the body-plasm of an individual with the body-plasm of his parents and ancestors, with all its conditioning of soul-activity or of mental characteristics and behaviour, is established fact; but continuity of the subjective thread or spiritual substratum is not observed fact. That the soul originates as the theory of traducianism asserts, and that the mystery of its becoming embodied is thus resolvable, is a view which the science of psychology renders untenable, if not grotesque. In so far as the noümenal man—in the phraseology of Kant—or the dominant monad—in the language of Leibniz—within the empirical self is concerned, psychology does not authorise us to say that man is the child of Nature, in the sense that he is wholly the product of evolution. The mystery of the soul's origin remains, and discontinuity, such as is annoying to the thorough-going evolutionist, must for the present, if not for ever, be in this instance recognised as having place within the evolutionary cosmos.

Another outcome of departmentality of thought that may be briefly mentioned is that the eminently learned in one sphere of knowledge sometimes make incursion into another sphere in which they are not equally at home, and are then apt to propagate erroneous beliefs. And those of us who, fortunately or unfortunately, are not in a position thus to mislead the public may, by otherwise similar procedure, mislead ourselves. Perhaps the chief offenders at the present hour in this respect are the mathematicians who expound the theory of relativity as a revelation of the structure of ultimate reality. They would persuade those of us who lack the skill to construct elaborate 'metrics' to see in space-time not one among other descriptive systems of relations conventionally devised *ad hoc*, but the naked 'stuff' which, alone wholly independent of our minds, dictates to them the phenomena they perceive. The specialists who would fashion contemporary philosophy in accordance with the scientific devices and policies of their day will probably always be with us. So also, perhaps, will theologians venturesome enough to steal their honey, when it is of the right flavour, for the theological hive. But the history of science forewarns us, and the philosophy of the sciences can forearm us, against confounding utility for the purposes of an abstract and special science with significance for the interpretation of knowledge, in its comprehensiveness, as to the world, in its concreteness.

We shall have further occasion to observe that whenever a certain field of facts is approached, in order to make a science of it, with assumptions, concepts, methods, etc. appropriate to some other science already enjoying a more advanced state of developement, instead of being allowed spontaneously to dictate those for which its peculiar data call, inadequate or misguided treatment is apt to be accorded to it. Of all the special sciences psychology has, perhaps, been the greatest sufferer from such departmental impertinences. When modern empirical psychology was in its infancy, irrelevant analogies, derived from the atomic theory which had proved illuminating in physics and chemistry, were thrust upon it, so that mental processes were studied as if psychology was a science of quasi-chemical laws. It had hardly recovered from this treatment when some of its friends, again hypnotised by the prestige of the physical sciences, sought to reform and simplify it by assimilating it to the sciences which make abstraction from the individual experient. It was overlooked that the standpoint of physical science, which deals with objects common to many subjects, presupposes that of psychology, and that the data of physics, as units of knowledge, presuppose the private percepts of individual subjects and are constructed out of them. So there arose a science described as psychology without a subject; and an attempt was made to despoil the science, which I shall later represent to be philosophically the fundamental science, not only of its privileged place, but of its very existence. Still more recently psychology has been handled as if it were akin to pure mathematics or to logic of the symbolic and computational kind. The notion of subjective activity has been called a superstition and a scandal; and that of an abiding ego has been declared superfluous, because replaceable by a logical construction such as a series of discrete entities. Description, in the sense in which a set of differential equations describe a machine, is said to be all that psychology should attempt in systematising its facts, while interpretative explanation should be eschewed as dangerously anthropomorphic. But a science of actualities, such as psychology, cannot be reduced to a science of ideal entities, such as mathematics, or to a science of abstract forms, such as extensional logic; nor can it be treated as a science to which the logistic method is adequate, or even adapted, without elimination of the more significant aspects of its subject-matter. As for anthropomorphism, I shall argue in a later lecture that nothing worth calling knowledge of the actual world can be had without some measure of it.

But psychology is not the only field of study that has suffered from incursions of the alien, or has been threatened with annexation by foreign powers. The mathematician and the mathematically minded philosopher, such as Kant, have doubted the claim of even physics, as distinct from dynamics, to be a science in the proper sense of the word; and mathematics has sought to honour physics by absorbing it into itself, and so giving it a place in the sun of genuine knowledge. Similarly, devotees of physical science have seemed suspicious of the claims of historians and sociologists. They have also, at times, refused the title of science to biology, so long as it retains concepts and principles other than those recognised in physics. Logic and theology, again, have each, in the days of their dominance, imposed a tyrannical rule upon the domains of the world of knowledge, proudly superior to unsatiated desire for possible worlds yet to conquer. The peculiar rôle of philosophy, as distinct from any and all of the special sciences, evades the apprehension of most persons who have not had the patience to make a serious study of it; so it is not surprising that, when philosophy urges its claim to be the *scientia scientiarum*, it provokes from the special sciences the question "who made thee a ruler over us?" Yet the instances which, at the risk of seeming somewhat rambling, I have deliberately accumulated in this preparatory lecture, with a view to indicating the insularity of departmental studies, should serve to manifest the desirability of a *scientia scientiarum*. And, in face of the rivalry and the tendency to anarchy that has appeared between the kingdoms of knowledge, there seems to be need of an over-lord who may indifferently minister justice between them; or at least of an "arbitrator from corruption free" who, by sympathy with their respective outlooks and endeavours, and by firmness in insisting on abatement of their pretensions, may bring them to co-operate towards a common end in co-ordinated knowledge instead of magnifying their own offices and trying to suppress others.

This task of mediation seems to be practicable; and the department of thought which should take it in hand may be called the philosophy of the sciences. I do not use this title with the meaning which it, or the similar name *scientia scientiarum*, has sometimes borne. I do not mean, e.g., a classification of the sciences according to their subject-matter, which is necessarily a somewhat arbitrary, artificial and provisional undertaking, having little significance. Nor do I mean a comprehensive survey of the sciences as they present themselves to the naked eye, so to say, or an attempt to find a sum-total of their generalisations, some of which, as they stand, may be of different orders of abstractness or of different degrees of approximation to metaphysical explanation.

Philosophy is indeed a synoptic view of things, an endeavour to grasp the main features of the world and to relate them as parts of one whole; and it is only thus that we can gain a sense of proportion and rightly estimate the significance of any particular science for our final conclusions about the world-process and the world-ground. But, just as an interpretation, and even an analysis, of any special department of knowledge, as it has organised itself, will not suffice to reveal its ultimate bearing upon our world-view, and must be supplemented by an inquiry into the presuppositions that may be involved in its very data and its categories, so must a philosophy of all the sciences concern itself with ulterior questions such as no science raises for itself. The sciences do not make it part of their business to explore their own presuppositions and to ascertain their own epistemological standing. They may proceed successfully in blissful ignorance that they have any presuppositions or that there are differences in what I am calling epistemological rank. And if they do estimate their own deliverances in respect of whether they express certain and necessary knowledge, probable belief, or pious opinion, positive fact or anthropic interpretation, and whether they possess finality or instrumentality, metaphysical import or lack of it, and so forth, philosophy cannot accept their self-estimations without scrutiny. Indeed, it is precisely in the sifting, in these respects, of the methods and products of the special sciences, that one of the chief and peculiar functions of what I would call a philosophy of the sciences consists. Far more important than a classification of the sciences and a comprehension of their unsifted deliverances would be an arrangement of the different departments of knowledge in order, according to dependence of one of them on another, and presupposition of one of them by another. Such a systematic arrangement of them is involved in what I understand by a philosophy of the sciences.

Unfortunately a difficulty here emerges. For presupposition and necessary conditioning are of more than one kind. There is the actual presupposition of fact by fact or happening; e.g. our awareness of temporal relations presupposes change in our field of sensory perception, and could not be forthcoming without previous acquaintance with change. There is also the logical presupposition of concept by concept; e.g. the explicit idea of change involves the logically prior idea of time. In the former case of priority we are following the *ordo cognoscendi*, the order of knowing and of knowability as determined by psychologically conditioned succession, or the dependence of stage upon stage in human cognition, as an actual, historical, process. This, it hardly need be said, has no bearing on the order of emergence of the specific kinds of fact which respectively form the primitive data of the several sciences, save to indicate that knowledge about abstract ideas is ultimately derived from, or occasioned by, acquaintance with concrete particulars. In the latter of the two kinds of presupposition which I have mentioned we are concerned with the order of thinkableness or conceivability, and with the question whether a concept is conceived "through itself" or "through another", to borrow Spinoza's phrases. In other words, we are dealing with degrees of analytical simplicity, abstractness, comprehensiveness, etc., in ideas. And this order, it does need to be asserted, is not necessarily identical, as Spinoza assumed it to be, with the *ordo essendi*—the only other order with which the *ordo cognoscendi* has been wont to be contrasted. If one may venture to coin a name for Spinoza's order, so as to be able to speak of it without begging the question of its identity with the order of being, I would call it the *ordo concipiendi*. One then indicates that abstract ideas or concepts, indispensable as they are for knowledge about actuality, are not themselves actualities and stages or agents in an actual process—as if time produced motion, and motion generated movables. We must not assume without inquiry, as philosophers commonly did until the eighteenth century, that pure thought is identical with knowledge of things, nor that thoughts or concepts are necessarily always concepts of existing, or actual, things; still less that they themselves are the ultimately real things. Hence the need of distinguishing the *ordo concipiendi* from both the *ordo cognoscendi* and the *ordo essendi*.

It is not always easy to keep distinct from each other questions as to priority in the respective spheres of causal order or temporal becoming, of knowledge-process or actual knowability, of timelessly logical or conceptual presupposition, and of metaphysical ultimateness. But it is important to try to do so, because what are first things in one of these orders may be last things in another. Thus, Spinoza's one substance, or God, is his *terminus a quo*, while in the order of knowledge it was reached by abstraction from previously known things and is a *terminus ad quem*. To recur to an example already used, perception of change is actually presupposed by our explicit apprehension of time, while the explicated concept of change involves or presupposes the prior conception of time. The psychophysicist regards a physical object or stimulus as the cause and actual presupposition of a given sensation, while the psychologist must regard knowledge concerning a stimulus as presupposing, and as occasioned by, reception of sensations. Common sense, impressed by facts concerning the dependence of psychical states upon the body, or by the scientific knowability of the body as compared with the mind, may regard matter as the fundamental reality; while the metaphysician, who by no manner of means can derive mind from inert matter, but finds no difficulty in conceiving matter as appearance of spirit, may regard spirit as metaphysically the prior. And there is no contradiction between the counter-assertions that are coupled together in each of these examples. Granted the standpoint and presuppositions from which each member of each pair issues, all of them may be true. For the orders are different, so that the statements in question do not meet and collide. All this shews the importance of taking account of standpoints, presuppositions, and diverse orders of priority. Neglect to do so may result in trying to establish equations between incommensurables, or in procedure like that of the jury which, taking down the various alleged dates on which the Mad Hatter began his tea, added them up and reduced the answer to shillings and pence.

The question then arises, which of the several orders of presupposition and dependence that I have enumerated should be chosen by a philosophy of the sciences for the purpose of determining the relation or the want of relation between different departments of knowledge, and of co-ordinating them, if possible, into one organic whole? If these orders were on a par in all respects that are relevant to the purpose, the only adequate or exhaustive plan would be to undertake a plurality of co-ordinations; to adopt each of these orders, one at a time, as a guiding thread, and follow the method which it dictates. If, on the other hand, one of these methods should be found to possess some quality in virtue of which it excels the others in any respect such e.g. as involving no concealed assumptions or foregone conclusions, that one should be given the preference. Now the inquiry that needs first to be undertaken by a philosophy of the sciences, and is of more importance than any particular issue, is that as to the logically and psychologically different processes and products that are embraced by the word 'knowledge', as it is commonly used, and which of them is chiefly encountered in the respective fields which we speak of as historical knowledge, scientific knowledge, mathematical knowledge, religious knowledge, etc. And, at least in so far as this main question is concerned, it seems to me that the method which pursues one of these orders of priority does possess an excellence of the kind that I have mentioned. It is that which adopts the *ordo cognoscendi*.

I therefore propose to use this order as the thread to follow in disentangling our knotted skein; and I believe that the method which may be described as analytic and genetic best enables us to trace those relations subsisting between the various departments of knowledge which are most important from the point of view of a philosophy of the sciences. This is the inquiry to which the rest of my course will be devoted. For the remainder of the present hour I will expound a little more fully my conception of what philosophy in general and a philosophy of the sciences are, and briefly set forth the reasons for my selection of the order of knowing for our guidance, rather than any of the alternative orders of priority that have been mentioned.

Philosophy, as I understand it, is not coextensive with science, as positivists of the nineteenth century represented; nor is it a prescientific occupation with the subject-matter of science, which advancing science supersedes, as some scientific writers have thought; nor is it to be restricted to thought such as must be equally applicable to all possible worlds, as some authorities on mathematical logic have maintained. Again, what I mean by philosophy is not baseless speculation, setting out from concepts fashioned according to individual predilection, though much of what has been produced, down the ages, by philosophers is perhaps largely of that nature. Philosophies, or particular systems of philosophy, have indeed been known by the names of their founders, and have been personally coloured. And there are some who consider this personal element to be essential to philosophy, as distinct from science, because philosophy should be concerned with life and not merely with logic; as if there should be as many philosophies as there are philosophers. A view somewhat akin to this was adopted by Dr Bosanquet, who affirmed that philosophy, "being, like language, art, and poetry, a product of the whole man, is a thing that would forfeit some of its essence if it were to lose its national quality".[1] But, as distinct from personal and national philosophies, there surely may be a pursuit and a method of philosophy, none the less involving the activity of the whole man that should be in all men, and dealing with every aspect of the world, of life, and of knowledge. I fail to see why there should not be a philosophy as cosmopolitan as is science, as free from individual bias or personal presupposition, and calculated to secure a consensus of experts such as is enjoyed by the sciences. The vagaries of philosophies are chiefly due, it would seem, to their deficient submission to the external control that may be exerted by forthcoming fact. Facts, as they are constituted at the level of experience-organisation that we call common sense, and as they are thence taken over by the several sciences, may indeed admit of analysis and of translation; but they nevertheless supply the only data for a philosophy concerning the actual world. They furnish the only touchstone by which analyses and theories may be tested. They dictate the method to be employed in philosophy, and render all beginnings from above, or from the *a priori* that

[1] *Proceedings of the Aristotelian Society*, N.S. vol. xv, p. 2.

claims to be independent of empirical facts, gratuitous guesses at truth which can only be distilled from our presumptive knowledge as it is systematised by the sciences. The restrictions upon the tendency to free composition that may be imposed by forthcoming knowledge, which is comparable with a passage set for translation, are severe. Recognition of them may circumscribe the region within which serious philosophy may move, and make the pursuit of philosophy seem relatively pedestrian. But if there are limits to the scope of our faculties and to the field for their exercise, it is wise to take account of them; and if there are rules of the philosophical game, it is becoming to play it in accordance with them. Philosophy, as thus conceived, is based upon the self-same data as are the sciences, and employs as many of the processes involved in the composite thing called the scientific method as the nature of its quest allows; but it adopts what I may call a different direction of departure from those data. The sciences, other than the psychology of cognition, each assume a theory of knowledge which it is not their business to sift and examine. They build with their data, or organise them, without raising previous questions as to issues prejudged in the mere acceptance of those data, whereas philosophy is concerned with the scientific exploration of those presuppositions which the more superficial departments of knowledge—if I may call them so without implying disrespect—take for granted. Thus the problems of philosophy are not the same as those of science. In exercising either its critical or its speculative rôle, philosophy can use no other material than that which the sciences supply; nevertheless it reinterprets, in the light of its own critical investigation of knowledge-processes, the facts and generalisations which the sciences yield.

A philosophy of the sciences will then begin with ascertaining what knowledge is, or what the various psychologically and logically distinguishable products which share the name of knowledge are composed of; whether they are characterised by certainty and by necessitation of this or of that kind, or by probability of this or of that kind; whether assumption or postulation or anthropic interpretation does or does not enter into their composition. Although it is not the business of philosophy to prescribe a method to any science, it is its business to examine all methods, and to ask what, if any, are the presuppositions of particular methods. A philosophy of the sciences will further inquire what categories are used by each of them; whether those that are used are indispensable, so that one science, such as biology, has a form of synthesis which is not wholly accounted for by a formulation in terms of the minimum of categories requisite for another science differing from it in generality, such as physics. Such a question raises for philosophy the further and more general question as to what explanation, in the various departments of knowledge, actually is, how many types of it there are, and what are their respective ranks from the point of view of logic. Verification, of which also there are several kinds and criteria, calls for similar treatment from philosophy. However, once a beginning has been made in the enumeration of the specific questions with which a philosophy of the sciences may busy itself, it is difficult to stop. Without mentioning more of them at this stage, I may observe that the office of the philosophy of the sciences with which we shall be mainly concerned in the present course of lectures is that of establishing some order among the chief types of knowledge, and some continuity in the transition from observation of the chaos of historical happenings, in all their multiplicity and fullness of qualitative content, to the pure and normative sciences on the one hand and to philosophy and theology on the other hand. In the latter departments of thought, if anywhere, we should look for that co-ordination of our knowledges which, as a kind of retrospective interpretation, consists in 'seeing whole' or in regarding the world as one unique thing; whereas the sciences which survey but dismembered parts and trace common strands are precluded from seeing what is called meaning, even if it be there.

Such being the conception of philosophy that is involved in what I would understand by a philosophy of the sciences, I may now observe that it leaves me with no choice but of the *ordo cognoscendi* for the purpose of inquiring into the nature of knowledge in general, of distinguishing the various processes and products that are accounted knowledge, of establishing relations of priority and dependence between the departments of knowledge, and of discussing their scope and validity.

The order of metaphysical priority, or of approximation to knowledge of the ontal, as distinguished from the phenomenal, is precluded, because metaphysics is a quest which is dependent on the sciences of the historical and the phenomenal. Metaphysical knowledge is the last kind to be attained, if it be attainable, and the approach to it is the most precarious. Not until we have passed, by way of the sciences, to metaphysics, can we set up a metaphysical system and speak of degrees of approximation to a metaphysical standard. This order, being conceivably knowable only after our pursuit is accomplished, cannot give us guidance while we are pursuing.

Similarly the logical order is ruled out. Philosophy of the actual cannot begin at the logical beginning because the logically prior, save perhaps a few empty forms, is only known from the logically posterior.

Though pure thought doubtless has its *a priori* conditions, most of the sciences which we are concerned to relate are not pure sciences but knowledges into the data of which the brute element of sense and its subjective effects enter. And there are no logical or *a priori* conditions of the forthcomingness and nature of sensory experience, or as to pure thought being applicable to sensible actuality. Again, there may be *a priori* or logical conditions of universal and necessary knowledge concerning actuality, such as Kant thought he found lying to hand in the Newtonian physics; but we now believe that no such knowledge finds a place in our natural sciences. Explorations of the logical presuppositions of knowledge, with which philosophers have sometimes occupied themselves, have thus been studies of a non-existent knowledge rather than profitable inquiries into the nature of such so-called knowledge as we have. And inasmuch as degrees of approximation to an ideal but non-existent standard are irrelevant to the more significant relations actually subsisting between the sciences, the logical order, or the *ordo concipiendi*, cannot supply the guiding thread of our inquiry.

Lastly, the *ordo essendi* cannot be chosen. In the first place we do not begin with knowledge of the *ratio essendi* of things. When we think we do, or proceed as if we did, know it, we are making assumptions with regard to the as yet undefined and unexplored process called 'knowledge'. We are dogmatically or uncritically taking for granted some theory of knowledge into the validity of which it is the function of a philosophy of the sciences to inquire. Philosophy must be presuppositionless in the sense of being aware of what its own inevitable presuppositions are, and of being alive to every assumption and act of daring venture that may be involved in the judgement that we know the *ratio essendi* of things. This self-knowledge, so to speak, on the part of philosophy, is only possible by means of a critical regress which consists in ascertaining what knowing actually is and how it has come to be what it is. The *ordo cognoscendi*, as I shall endeavour more fully to shew in my next lecture, is the only way by which it is possible for us to arrive at a known, as distinct from a conjectured, *ordo essendi*. Hence the former order alone can guide our discussion of the less superficial relations subsisting between the different departments of knowledge, all of which are concerned to impose a rational form upon concrete actuality with a view to understanding it, or else to discover the rules which thought should follow in order to understand—i.e. to regard, in so far as it is possible, as necessarily connected—what is first given to experience as if contingent.

Lecture II

THE RELATION OF THE PSYCHOLOGY OF KNOWLEDGE TO PHILOSOPHY OF THE SCIENCES

THE FIRST of all the tasks to be undertaken by a philosophy of the sciences is to explicate what knowledge, or knowing, itself precisely is. Without raising this fundamental question any attempt to discuss the relations between the different departments of knowledge, and to arrange them in a philosophically significant order, would be but superficial. Whether as *scientia scientiarum*, in the sense in which I have used that name, or as metaphysics, philosophy presupposes, or begins in, the science or theory of knowledge. Any science which studies what goes on outside our minds must be contemplated in connexion with what goes on within our minds, if it is to possess significance from the point of view of the philosopher; for he is concerned to see things whole. Philosophy must, therefore, analyse knowledge-processes into their simplest terms and examine the faculty of knowledge which is involved in all knowing. If a critical examination of this faculty admits of being successfully carried out, it will lead to the establishment of some general principles which can be used for determining the scope, rank, and relations of our various knowledges. And these principles will be such that they will not be affected by further advances in departmental fields or by changes of view or theory concerning any particular kinds of known objects.

The phrase "departments of knowledge" would seem to imply the belief that there is only one process and only one generic product denoted by the word 'knowledge', the same in all the fields which we call sciences, and that the connotation of the word 'knowledge' is so unambiguous and clear as to be transparent to everyone and to call for no examination. But let even a person who has never studied a text-book on psychology be asked whether the word 'know' suggests to him precisely the same mental processes and the same shades and kinds of certainty as it occurs in the following propositions: I know that it is now light; I know Mr *X*; I know that 2 and 2 are 4; I know the Lord's Prayer by heart; I do not know Arabic; I know that King John signed the great charter; I know that on such and such a date there will be an eclipse; "I know that my Redeemer liveth". Probably such a person, after a moment's reflection, will find occasion to suspect that when we glibly use the word 'knowledge' as if it always bore exactly the same meaning we are guilty of terminological vagueness. Yet some philosophers, when they have passed judgement on dogmatism, scepticism, and criticism, or have discussed the faculty of knowledge and the conditions of knowability, seem often to have made an assumption similar to that of the uninquisitive man. I mean that they have taken it for granted that at least some of our forthcoming knowledge of the actual world is characterised by necessity and certainty, and in that respect is distinguishable from belief or opinion. It is often said that there must be some knowledge, in this sense, in the possession even of the sceptic whose maxim is "he who shuns certitude is sure", else he could put no disconcerting question, affirm no doubt, make no denial, and draw no distinction between truth and error. Indeed one purpose of the inquiry into the preconditions of knowledge has been to discover what the most thoroughgoing sceptic is logically compelled to confess that he necessarily knows. I ventured to assert, in my last lecture, that the certain knowledge of the actual world, the *a priori* conditions of which have been investigated by philosophers of one particular school, does not exist; and I can now imagine the question arising in the minds of some of my hearers, "how do you know that it does not exist, and how can you say so without tacitly claiming some item of knowledge of the self-same kind?" Or, to put the issue in a more general form, is it not impossible for knowledge to examine and criticise itself, since for that very examination certain knowledge and a criterion of truth or a canon of criticism are already presupposed? Such a question indicates that there are difficulties to be overcome if I am to escape being set down as a sophist when I proceed to affirm that a critical regress is possible which does not involve infallible knowledge in its examination of what purports to be knowledge, or when I deny the forthcomingness of that kind of knowledge which my adversaries claim for themselves and would impute to me as a necessary fixture in my mind, which I have overlooked in making the inventory of my mental furniture.

I will pursue in a later lecture what I deem to be the only kind of critical regress, or examination of knowledge itself, that is both possible and capable of yielding philosophically significant information—i.e. information as to such so-called knowledge as we actually have. But I may say at once that this examination is essentially psychological, and therefore not of the same nature as the classic example provided by Kant. And I shall have advanced one step towards removing the suspicion that such a psychological examination is foredoomed to futility if I can brush aside one commonly adduced objection to the psychological method of procedure. This objection may be valid as against Locke's psychological inquiry, in contrast with which Kant's critical investigation of the faculty of knowledge is wont to be described and extolled by his disciples. But it is a mistake to regard the procedure of Locke, a pioneer in this field of research, as the only type which the psychological method can follow; as e.g. when Edward Caird, in his work on Kant, speaks of "the Lockian or psychological theory of knowledge". Locke worked with assumptions which psychology has since discarded. He assumed, e.g. that the objects of sensory perception are states of the percipient subjects; and he wrote before the all-important distinction between individual or unsocialised experience and common or over-individual experience was discerned. Consequently, as

Caird observes, Locke bids us examine our faculties in order that we may discover the nature and limits of our knowledge, "very much as we might examine a telescope in order to discover whether there was any flaw in its construction which might distort our vision of the objects seen through it". And, Caird continues, "if this were the sense in which Kant bade us criticise our faculty of knowledge, it would not be unreasonable to object that we cannot examine the mind except with the mind, and that any defect in the mind which might hinder us from knowing other objects or from knowing them correctly, would equally hinder us from knowing the mind itself". "To this", he concludes, "from a purely psychological point of view, there seems to be no answer."[1] From a purely Lockian point of view, perhaps, there could be no answer. But it is another question whether any answer can be given from a psychological point of view, with the eye open to Locke's defects and errors.

A relatively small obstruction may thus be removed from my path, but a more serious one still remains. It is said that in general the criticism of knowledge, or knowing, presupposes some certain and universal knowledge; and the saying has seemed to many philosophers to be beyond the possibility of refutation. If all that it means is that, in so far as knowledge involves thought, the laws of thought on which logic rests are preconditions of knowledge, this statement cannot be gainsaid. But it generally means more than that; viz. that knowledge

[1] *The Critical Philosophy of Immanuel Kant*, 1889, vol. I, p. 10.

has other *a priori* conditions also. And, indeed, if knowledge is characterised by universality and necessity it must have further *a priori* conditions. For logic is not the only apparatus involved in knowledge of actuality. It does not supply us with a single true or valid premiss; and thought is not identical or coextensive with knowledge, though it is an essential factor of knowledge. But, while it must be allowed that knowledge has logical conditions, it is open to us to doubt whether there is any universal and necessary knowledge, requiring *a priori* elements other than the laws of thought, and whether a modicum of such alleged universal and necessary knowledge is an indispensable condition of the forthcomingness of those actual products which we are pleased to call knowledge. The ancient sceptic, such as Sextus Empiricus, was no negative dogmatist; he did not claim greater validity for his own sceptical assertions than he would allow to the dogmatic assertions of others, but merely maintained that all assertions are more or less uncertain. And in a somewhat similar spirit I would submit that the proposition which we are able to assert without possibility of refutation is not that there must be some certain knowledge which even the sceptic presupposes, but that we possess presumptive knowledge, the ultimate presuppositions of which may in turn be found to be presumptive. To avoid misunderstanding I should observe that throughout the present context I am using the word 'knowledge' as solely denoting thought valid of actuality, and not as embracing also truth about the purely ideal. And to suggest that all our knowledge of the former kind may be presumptive is but to decline to accept as self-evidently applicable to it the conception of knowledge which was entertained by Kant and rationalistic philosophers. Our knowledge may turn out, on examination, not to consist wholly of elements characterised by necessity and universality, but only of ingredients having that nature along with ingredients of other nature. The mistake, as it seems to me, of those who pursue epistemological inquiry on the Kantian lines rather than on those imperfectly and somewhat erroneously sketched by Locke, is that they take presumptive knowledge at its face value, or interpret it in terms of a preconception, and then proceed to inquire into its *a priori* conditions.

The phrase *a priori* has come to possess several different meanings; but in the present connexion it usually denotes "that which is true, no matter what". One representative of the method of critical regress which pursues the *ordo concipiendi* has recently expounded the phrase precisely in those terms; and he identifies the *a priori* and necessary truth which all knowledge is supposed to involve with that, the free recognition of which is compelled by "the conditions of intelligible communication". This intelligible communication, the actuality of which sets all the problems of philosophy, may, however, be but what I have called presumptive knowledge; and so it should be called until it shall have substantiated its claim to be anything more dignified. But that the conditions of intelligible communication, or of the existence of our sciences, are absolute and universal truths in addition to the fundamental laws of thought, and that these truths must obtain "no matter what", seems to be an assumption requiring examination rather than a self-evident axiom. Another representative of this method of conducting the critical regress, which is supposed to presuppose some recognition of universal truth, tells us more explicitly that "the name *a priori* can be applied only to those elements of truth which are presupposed in all consciousness of objects": elements which are consequently incapable of being treated as on a par with special facts or laws because they are principles through which all truth must be seen. This statement is typical of a school, rather than expressive of an individual opinion; but it is somewhat indefinite because of the ambiguity of the word 'objects'. If "consciousness of objects" means private apprehension of the simplest percepts or sense-data, such as a patch of colour, that apprehension presupposes no logically necessary truth-principles: it is characterised only by actual necessitation. But probably by 'objects' are meant the things of public or common experience, between which science traces causal relations: in which case it is disputable whether the presuppositions of their forthcomingness, or of their knowableness, are necessary principles such as must hold, "no matter what", or are human postulations and devices *ad hoc*, whose justification as knowledge-principles consists in their practical successfulness and in their providing us with a scientific kind of experience which we are under no logically dictated necessity to seek.

Thus the obstruction which threatened to block my proposed empirical and psychological procedure, at first sight seeming to be a self-evident truth, turns out to be rather a dogmatic assumption. In order to make a critical examination of presumptive knowledge it is not necessary to assume the existence, and much less the necessity, of a definitely conceived and defined kind of knowledge, whose certainty is self-guaranteed and which must be used in the critical sifting of all other knowledge. Our presumptive knowledge can turn its perhaps imperfect eye upon itself, as a process of knowing, as well as upon its objects and its products. It is not hindered from doing so by the possibility that, over and above some elements of certainty, in virtue of which it passes for knowledge, it contains venturesome postulates and generally cherished, but not logically certified, opinions. Our inquiry may clarify our first vague notion of knowledge and yet presuppose nothing but the fact that, by performing certain mental operations on confused opinions, clearer conclusions emerge which are stable beliefs capable of commanding general assent. The certainty which in the past has been asserted to be essential for all that can be called knowledge, and to be presupposed in all criticism of alleged knowledge, can be replaced by the security or high probability of belief which passes for knowledge because of its serviceableness in enabling us to cope with life: which serviceableness bespeaks some relationship, however indirect and subjectively vitiated, with the ultimate truth of things.

Thus the procedure that would seem to be prescribed to us by the fact that there is such a thing as intelligible communication between human beings is to set out from our presumptive knowledge without any foregone conclusion as to what knowledge ought to be, conceivably may be, or necessarily must be; to isolate those elements in it which constitute the fundamental certainties which simply are, and which may of themselves not constitute, or be allowed the name of, knowledge, and to sift out the uncertainties or the venturesome elements that unavoidably enter into what we commonly account knowledge. We may thus become explicitly aware of any presuppositions which presumptive knowledge may contain, and be able to allow for them. It is indeed often said to be as foolish to ask how knowing can know as to ask how existent things came to exist; but how knowers come to know or to possess presumptive knowledge, and what their knowing then actually consists in, are questions for the answering of which positive data lie to hand for exposition and analysis. Thus the logical preconditions of what some philosophers may wish knowledge to be are one thing, and the actual preconditions of such so-called knowledge as we actually have, and perhaps only can have, are another thing. It is the latter quest alone that seems to be worth prosecuting. The need to prosecute it is obvious. For the 'instinctive' philosophising of mankind, or the organisation of experience for the practical purposes of life, to which thought is instrumental, has created many perplexing problems for the critical philosophising of cultured or sophisticated men. Our very language, which is an indispensable tool for thinking, embodies foregone conclusions which exclude what we can now see to be theoretically possible alternatives, making these seem to be absurd to common sense. Common sense and its primitive metaphysic thus require overhauling before a critical philosophy of the sciences can be constructed.

As for the method to be employed in this necessary regress, I would repeat that the structure and nature of our knowledge can only be ascertained by analysing it, ascertaining its sources, and seeing how its ingredients have actually been compounded, or with what mental cement, so to say, its objective data have been compacted. It is only by tracing the developement of the knowledge-process, from its simplest forms and from the earliest stages of it that are now discoverable, to its culmination in the advanced sciences, that the nature of its finished products and the scope and limitations of its processes can be learned. These statements supply the explanation of my previous assertion that the *ordo cognoscendi* is the sole route that can possibly lead to a known, as distinct from a conjectured or assumed, *ordo essendi*. They also imply that the psychology of cognition, analytic and genetic, is the fundamental science from the point of view of a philosophy of the sciences. In submitting this judgement I lay the foundation-stone of the philosophy of the sciences which is to be constructed in my subsequent lectures, and present my first conclusion as to the relation in which any one department of knowledge stands to others.

The statement to which I have been led, that psychology concerned with cognition is to be exalted to the place of the first propædeutic to philosophy is open to a misconception against which it behoves me to guard. For I do not mean by it merely that the student of philosophy would do well to read some psychology before he approaches the literature of metaphysics or other branches of philosophy; I rather indicate the systematic order in which I take psychology to be related to philosophy and the other sciences. And this view may be challenged on the ground that psychology, such as would be more than an inventory of observed facts, itself involves notions and ways of thinking which need critical scrutiny as much as do the notions and ways of thinking employed in any other science which supplies data to the philosopher. This is to be admitted: it is not on account of any imagined superiority of psychology, in the respect just mentioned, that I claim priority for it. Psychology is a body of presumptive knowledge like any other science of the actual, and, like any other science, needs to submit the presumptions involved in its very vocabulary and grammar to examination. What entitles it to its position of philosophical priority is that the facts which it supplies, unlike the facts supplied by any other science, elucidate the origin, nature, scope, and limitations of the notions and ways of thinking which psychology itself, as well as other presumptive knowledge, employs. It seems to me that there neither is nor can be any independent and still more fundamental department of philosophy, which, in the order of knowing, psychology presupposes; though strenuous thinkers, such as Husserl, still endeavour to conceive of one. Husserl's "phenomenology", purporting to be the basal science propædeutic to all the sciences, is a pure and *a priori* doctrine of experiences "in their essential aspect" or of consciousness as independent of a world or of Nature. It deals with essences known independently of perception of things, and invokes an *Evidenz* which is a seeing of these essences and their truth through a pure, 'categorial', intuition; but whether such intuition is as much a non-actual abstraction as the imaginary number or is an actual component of actual experience would seem to be a question which admits of an answer only by appeal to psychology, or to what we know about experience and about analysis of it as compared with abstractions from it. And a similar objection may be urged against all the epistemologies of which Husserl's is a typical representative.

Psychology, then, I repeat, alone elucidates the ways of thinking which it, as much as any other department of presumptive knowledge, employs. But whereas the other sciences stand in need of a sifting, as to their concepts and modes of thought, which they do not and cannot themselves undertake, psychology comprises the investigation, and supplies the means of investigation, of the concepts and modes of thinking of which it itself needs, provisionally in the first instance, to make use. Using the *ordo cognoscendi* it unfolds the genesis of what the *ordo concipiendi* takes for first things. And, as I shall presently argue, the question of the origin of knowledge is relevant to the question of the validity of this or that kind of knowledge. It is more obviously essential for ascertaining the nature of knowledge, which otherwise is very liable to be defined in such a way as to indicate a class of which there happens to be no member.

But instead of prolonging this abstract discussion of generalities, which to some may seem indefinite and unenlightening, I will proceed to set forth in the concrete how I understand psychology to fulfil the high office with which I have credited it.

And first I would meet more precisely and concretely the objection that psychology is debarred from possessing priority in the system of knowledges by the fact that, no less than any other department of knowledge, it involves obscure and assumptional conceptions. This, it will be remembered, is the reason why many philosophers look elsewhere for the first propædeutic to philosophy, viz. to a transcendental logic, or a theory of knowledge which is neither psychology, nor logic, nor an amalgamation of the two, but a *tertium quid*, a science *sui generis*: a science which I have ventured to describe as concerned with a non-existent ideal of knowledge rather than with any actual product upon which we are wont to bestow the name of knowledge. The all-important fact about our presumptive knowledge is its forthcomingness. It may be assumptional, impure, inadequate, and in need of translation. But both as a body of data and as an intellectual tool for systematising fact-data it is in the first instance the only knowledge or accepted belief that we have. All that we can do is to make the best of it. It is our sole external control, and the only touchstone by which we can distinguish science from groundless speculation. It contains our *explicandum* though it may be a faulty explication of it. It cannot be scrutinised from without, because we cannot get outside it in order to look down upon it. But fortunately it can be scrutinised from within itself. If, in the abstract, this statement seems paradoxical, the history of science and philosophy shew that in some measure similar problems have been solved *ambulando*. For instance, science treats, in its initial stage, of material coloured bodies operated upon in space and time by forces. It proceeds to eliminate colour as a subjectively conditioned and secondary quality, to dispense with forces as imaginary, and to resolve space and time into differentiations of space-time. If this be an advance towards purer and more adequate knowledge of reality, or from obscure and figurative to clear and literal representations, the criticism of previous notions involved in it has been due to science's thinking along her own lines and not under the guidance of any deliberately adopted theory of knowledge. Moreover, however far scientific knowledge goes in such self-improvement, its theory is verified by experimental appeal to the crude appearance-facts from which it sprang.

The case is the same with psychology. And whether the assumptional elements involved in common sense and empirical science be supersedible or not, they need not be regarded as meanwhile radically misleading. Our equations, so to say, will not be affected by a change of origin, co-ordinates, or symbols, if we become able to make such changes. In other words, an existing stabilised science such as empirical psychology, for all its initial defectiveness from the point of view of a subsequently reached theory of knowledge, must be some version or function of the purest truth about mental actualities, by which we would fain replace it; else it could hardly have come to being and persistence. And after all it does not matter whether truth mirrors actuality as a lake reflects the surrounding hills if, like a map, it is relevant to the country through which we are travelling. The psychology of to-day, then, will be truth relevant to actual knowledge-processes, and will convey some sort of knowledge about them, even if its thought-apparatus is here and there provisional or assumptive and its concepts are figurative rather than literal. It is undoubtedly knowledge in the sense of not being illusion or fancy. But what exactly any such knowledge is, and to what extent it is characterised by necessity and universality, by certainty or by probability, by subjectively uncontaminated objectivity or by interpretativeness, is precisely the first question which a philosophy of the sciences must consider. And it is questions of this kind which psychology, and psychology alone, save for some assistance from logic, can answer. Hence I regard it as the fundamental science.

Postponing these questions for the present, I would announce that the psychological procedure which will be followed in dealing with them is to be partly analytic and partly genetic. Unless the one of these methods is controlled by the other, either is liable to lapse eventually into the pursuit of mythological conjecture. For instance, the exclusively genetic doctrine of 'pure experience', as taught by Bradley or James, resorts to that endless genealogising which long ago received apostolic censure; and the exclusively analytic doctrine of pure sense-data and unsensed sensibles, in terms of which cognition has been described by another school, treats of fictions which have escaped the control of facts. And I may remark that neither the use of the genetic method at all, nor the combining of it with the analytical method, need involve any confusion of the two. The two methods can be pursued together without any false identification of analytica with antecedents, i.e. of distinguishable but perhaps inseparable simple factors with earlier events or stages. Nor, again, is genetic psychology committed to sensationism or associationism. Such opinions, expressed by critics of the genetic method, imply a misunderstanding of its nature and have force only against abuses of it. But more formidable is the declaration that inquiries into the origin and developement of the knowledge-process, so far from revealing anything of philosophical significance about knowledge, are irrelevant both to the nature and the validity of knowledge. Besides being important in itself, or as a fundamental issue in general epistemology, this issue is of moment in that it meets us recurrently in its applications to the sciences severally. It is therefore necessary to discuss it and to vindicate the genetic procedure which I propose to follow.

The most general form taken by the objection which we are to weigh is the assertion that psychological conditionings have nothing to do with logical presuppositions, or with validity, and that the truth of any proposition is wholly independent of how the proposition came to be known or believed. And certainly propositions which we regard as true have sometimes come to command belief and to be asserted on inadequate grounds, in consequence of fallacious reasoning and through alogical causes such as what is called instinctive association. Yet this fact does not serve to establish the general statement that the truth of a proposition is wholly independent of its origination. Indeed this dictum, in the sense in which it is true, seems to have little significance and, in the sense in which it would be significant, to be untrue. For it ignores the difference between truth which is known to be true—the only kind that is usually and properly called truth—and what is said to be truth though it is not known to be true or is independent of our truth-recognition. The latter kind of truth is rather fact than correspondence of ideas with fact, or being, rather than relevance of propositions to being. So long as this so-called truth remains unrecognised, or is not known to be true, its subsistence and its non-subsistence are alike nothing to us. It is no use there being true propositions if we do not know them to be true. Our sciences are not composed of such truths. They are composed of propositions concerned with actual things and believed to be valid of things. And in order to know any such proposition to be true we do generally need to have ascertained how it was arrived at and how it has become known to be true. Thus we may allow that a judgement, once it has been established, is true, no matter how we came in the first instance to be persuaded of its truth; but if we ask ourselves on what grounds an alleged truth concerning actuality has been asserted, and what right we have to regard it as true, we raise the psychological and genetic question, how our knowledge came to be what it is; and we are no longer concerned with pure logic alone. If we abstain from asking this question, we are claiming to possess truth without being able to justify our claim, and are playing the dogmatist.

Dogmatism of this kind seems to me to lurk in all realistic theories of knowledge, from the common-sense realism of Reid to the logical realism of Plato. These theories treat what is, or conceivably may be, the final product of an actual and tentative process as if it were a static, heaven-descended, prior reality, so sacrosanct as to render inquiry as to its suspected earthbornness and its humble antecedents impious. If the realistic or the rationalistic theory of knowledge were true, genetic inquiries would certainly be irrelevant to the science of knowledge; but I hope to shew later that empirical criticism dissolves the knowledge which that kind of theory sets up, or presupposes, into an ideal comparable to the line without breadth. Meanwhile some weak points in the arguments for the rejection of genetic studies as futile may be exposed.

It will probably be allowed by those who repudiate the genetic method in epistemology that it supplies us with a more or less true account of the historical process by which knowledge of scientific truth has been humanly acquired. It is admitted that such truth, from the point of view of history, is the final stage of a process of trial and error, and of abstraction and synthesis, in which categories and hypotheses were invented, selected and refined. Now, to take one such stage, viz. the finished product as it is at present constituted, as so different in nature from all the preceding stages that it alone is to be regarded as truth or knowledge while they are to be looked upon merely as causes of its coming to be known, seems arbitrary. There would seem to be an unaccountable discontinuity between the earlier stages of an apparently continuous process, in which postulation, interpretativeness, and pragmatic verification are to the fore, and the final stage which is supposed to stand for absolute truth bearing logical certification. We may ask why the earlier stages are but irrelevant events and the last is fraught with truth-import. In view of this appearance of arbitrariness it is incumbent on upholders of a non-genetic theory to shew that at the last stage of conception, or in a finished product of knowledge, all the instrumentality and preparatoriness of earlier stages of the process have been strained out, that postulatory categories have been replaced by formal categories or *a priori* factors, and so on. And as it is not possible to do this in piecemeal fashion the short way is generally adopted of asserting that the truth embodied in the final stage or product is apprehended or constructed in an entirely new manner, viz. with immediacy. The non-genetic theories of knowledge can only justify themselves by freely invoking such immediate apprehension; but analytic psychology suggests doubts as to its real immediacy, and consequently threatens the foundation on which these theories are based.

That immediacy needs to be invoked by opponents of the genetic method strikes any reader of either their strictures or their self-defences. A typical example may be adduced from Sidgwick's criticism of the historical method as applied in the theory of knowledge. In several passages of his work entitled *The Scope and Relations of Philosophy* Sidgwick refers to the superfluousness of the intuitionist's claim that certain ideas or beliefs are innate. An apprehension such as, e.g. that of the oughtness of an act, he observes, may be intuitive without being innate. And he ascribes intuitiveness or immediateness of apprehension to various acts of cognition which are neither innate nor analytically simple. One such act is what some would call our perception, and what he calls our notion, of common space. "Space", he says, "does not mean to me successive feelings of movement, conceived as simultaneous from association with simultaneous feelings of touch, though this may describe the process by which I have come to have my notion of space." One may doubt whether 'space' means presentations of movement, etc. to any genetic epistemologist; but, letting that pass, Sidgwick insists that the notion of space which we have come to possess is none the less that notion of space because analysis has revealed its conditions, antecedents, and concomitants. And one may acquiesce in his distinction of the conception of space from the motor-presentations, etc. which psychologically condition our attainment of it. But the important question is whether these conditions are exhaustively and correctly described by Sidgwick as merely antecedent and concomitant events, epistemologically irrelevant to the nature of our conception of space. We may ask whether that conception is not just a stage further, not indeed in motor-sensation, but in the subjective correlation of private touch-space with movement-space, and of private spaces with one another to constitute public space, at which we conventionally stop because it happens to be sufficiently convenient for the purpose of common-sense thought and practical life. In the former case, to which Sidgwick seems to assume there is no alternative, the psychological antecedents would be comparable to steps from a subterranean cave to the surface of the ground, by ascending which one may come to perceive the sunlight, but which do not figure in the chain of physical and psychological causes which mediate and constitute the perception of light. For him there is a discontinuity and a disparateness between the psychological conditionings of spatial intuition and the conception of common three-dimensional space: a disparateness similar to that between climbing the dark steps and suddenly beholding the sunshine. It follows that space-perception, or space-conception, according to this view of it, is an inrush of wholly new and unique insight, arriving in each individual opportunely or in the fullness of time, and that it is performed, like sense-impression, in a flash, or is as immediate as if it had been innate. It must also be performed alike by all common-sense persons. And this seems mysterious, especially when we bear two considerations in mind. The first of these is that space is neither a simple relation, like difference, nor a universal, like whiteness, that is distilled from sense-particulars, any more than it is an impression or a simple percept. The second point is that private spaces, or the antecedent elaboration of spatial relations between percepts, accomplished by each individual for himself, are diverse though capable of being correlated. Moreover, we should expect a notion that was so generally shared and formed with such immediateness to put us in touch with relations independent of our subjectivity and to command lasting recognition and unalterable description from science through all stages of its advance. Yet this is not the case. The common-sense notion of space, alleged to be read off with immediateness or to be fashioned of necessity, does not now serve some purposes of science so well as one that is essentially different. Thus the theory which assumes the psychological conditionings and the earlier stages of space-intuition to be extraneous events, instrumental to it but not constitutive of it, raises difficulties. These are resolved, on the other hand, if we regard space-conception as of the same nature as the constructive processes with which it is historically continuous, and as differing from them only in that it is adequate for purposes of common life for which they are inadequate. Sidgwick rightly saw disparateness between motor-sensations and the conception of space. But genetic psychology is not nowadays committed to sensationalism: it speaks of subjective activities such as correlation and abstraction, and of stages in conceiving prior to arrival at the finished concept. These activities begin in individual experience and are further developed through social intercourse. Ignoring this fact, Sidgwick did not refute the genetic theory of knowledge, but only sensationalism. He also overlooked that genesis includes epigenesis, or the growth out of something into something else, in which the nature of the 'something else' is partly determined by the 'something'; and consequently his strictures apply to what has been called the naturalist's fallacy, but not to the genetic method.

Distinguishing between these things which differ, we may now say that the issue narrows down to the ultimate question of immediacy: antecedent stages are mediated, but the final act of knowing is immediate apprehension. This question has been rendered obscure by the ambiguity of the word 'immediate'. We can distinguish two standpoints from which immediacy may be asserted, and consequently two kinds of immediacy. At the moment of perceiving a thing, say a sparrow on the house-top, we are unconscious of performing, or of ever having performed, synthetic mental operations relevant to perceiving the sparrow. From the standpoint of the experient or the experience at that moment the perceiving is immediate, in that the percept has the unity and simplicity of a product of instantaneous photography. But such immediacy is, from another point of view, only unawareness, at the time, of actual mediations. For when, as psychologists, we reflect upon our perfected perception of the sparrow, we find that it could not have been enjoyed without previous experiences, lessons learned from them, and mental dispositions previously established by them. The facility with which we now perform the complex act of perceiving a sparrow has been acquired. And if the genetic account of perception submitted by Ward, who emphasises the act of combining impressions with residua, images, etc., may appear to differ somewhat from that more recently furnished by the school of *Gestalt*-psychology, the diversity as to detail does not throw suspicion on the assertions that our facility is acquired and that our perception is immediate only in the temporal sense of being rapid. For if, as the *Gestalt*-psychologists tell us, there are segregated wholes, or parts immediately given as belonging together rather than subjectively put together, in a sensory field, they also tell us that such segregated wholes or *Gestalten* in most cases do not correspond to those which we call physical objects or things, such as a sparrow. The *Gestalt*-psychology may knock another nail in the coffin of sensationalism, but it is compatible with the genetic theory of knowledge.

We may conclude, then, that just as a *virtuoso*'s easy performance of a difficult piece of music bespeaks neither his composition of it nor his ability to have performed it with the same consummate skill at first sight, so what is called the immediacy of our adult perception and conception may bespeak neither innateness nor unacquired facility. A theory of knowledge that takes artifacts such as the concept of space, in virtue of their familiarity, for pure and immediately apprehended data, seems to confound the two standpoints, and the two kinds of immediacy, which I have distinguished.

Further ambiguity attaches to the word 'immediacy' in that it is used both in a psychological sense, when absence of causal linkages is indicated, and also in a logical sense, when 'immediate truth' means uninferred truth, or truth which calls for no proof because it is self-evident. We need not now examine various fallacies which have been generated in various departments of knowledge by this double usage of one word, such as the common confounding of psychological with logical certainty, or of pragmatic with logical certification: for only the one of these kinds of immediacy, viz. the logical, is relevant to our present inquiry. That is concerned with the self-evident validity of the final products of knowledge-processes and their independence of prior stages, in those processes. These final products for which immediacy is claimed comprise concepts, such as that of space, and propositions, such as what are called axioms.

With regard to concepts, perhaps enough has been already said to shew that the immediacy with which they are now fashioned or apprehended by us *may be* only of that spurious kind which consists in unawareness of actual mediations. And the fact-controlled psychology that traces stages in all conception which, as in the case of space-conception, is more complicated than the simple extraction of universals out of particulars, will compel us to substitute 'is' for 'may be', unless the rival theory of knowledge can prove that immediacy of the genuine kind is involved. But I shall presently argue, when considering the immediacy of propositions, that even this genuine immediacy, established by psychological reflection rather than assumed from the unreflective standpoint, presents no effectual bar to the further prosecution of genetic inquiry. Meanwhile I may refer to two views, concerning synthetic concepts and the scope of their usage, often implied in non-genetic theories of knowledge, both of which are convicted by the psychology of cognition of being dogmatic or groundless.

The one is that conception is analogous to perception in being an equally direct apprehension of an object, and of an object which in spite of not being perceptual or impressional none the less has the same actuality as that for which an impressional core is our primary, and indeed our only direct, vouch. It is a tenet of common-sense realism that conception is of this nature, and the belief is sustained by our usage of the phrase 'concept of'. This usage is misleading. The words 'the concept of space' naturally suggest that there is an indubitably actual or real, if not a substantial, entity called 'space', and that we have a conception of it just as we may have a mental image of an actual house. But though the word 'conception' may be allowed to stand for apprehension of an object in addition to the previous constitution or derivation of it, that ideal object cannot, without further ado, be said to have an actual counterpart. We have a concept of a line without breadth, for instance, which no one takes for an actual existent. And if, instead of the expression 'concept of space' we used the word 'space-concept', which leaves open whether or not there is an actuality corresponding to the concept, the dogmatic metaphysics involved in the commoner phrase would lack its linguistic support. A concept such as that of space denotes what we *think* exists, and think in order to systematise our perceptions of what does exist. It is adopted as valid if things are found to behave *as if* the ideal object were also an actuality. Such, from the point of view of psychology, is the cognitive status of concepts of that kind.

The other view which psychology can shew to be baseless is that concepts, once they have been obtained, are independent of the contexts in which they were suggested and fashioned, and of the subjective agencies involved in their fashioning. Certainly the logical connexion of ideas is independent of the mode of origination of the ideas; but it is not self-evident that the scope of their *knowable* applicability and the range of their significance are other than coextensive with the experience-contexts within which they were evoked, and from which, by abstraction, etc., they were derived. Again, as to their subjective conditioning, it by no means follows that concepts which are necessarily of over-individual, or social, origination, so as to be independent of the idiosyncrasies of any individual, are over-social or absolute in respect of origination and relevance, so as to be independent of the mentality of all mankind. To assert that they are is a dogmatic leap beyond what the science of cognition has established.

To turn now to immediately known propositions, and to come to the final question of the relevance of even genuine immediacy to truth or certainty.

The process which establishes the fundamental principles of logical inference, on which logic and mathematics depend, but which are not premisses out of which truth is deduced, is that known as intuitive induction. It provides other truth, characterised by immediacy of the genuine kind, that is of sensory or experiential origination: e.g. the judgement (from a single instance) that in all cases an equilateral triangle is also equiangular. These latter intuitive inductions occupy a unique place in our knowledge. They alone yield truth possessing all of the following characteristics: derivation from the impressional or perceptual, apprehensibility with genuine immediacy, objective necessitation, and universal validity. They are sometimes called *a priori*; and if we are pleased to call them so on account of their universality and immediacy, they would seem to be the only *a priori* truth forthcoming that is directly and necessarily applicable to actualities as contrasted with the entities with which the pure sciences deal. It is the immediacy, of which they present us with an indisputable instance, that is our present concern. And it is the supposed absoluteness of this immediacy that causes immediate or self-evident truth to be regarded as superior in certainty to all inferential or mediated truth. But even this immediateness is not to escape our scrutiny in respect of its possible genesis, nor to receive an uncritical beatitude in respect of its assumed absoluteness. The immediacy of intuitive induction, like that of impression and apprehension of simple relations such as difference, consists essentially in a temporal quality: it is intuition in a flash. But such rapidity of apprehension may be the outcome of practice, or of quick imaginal experimenting and inference, or of subliminal synthetic processes. No intuitions can be literally instantaneous; any flash must be of finite duration: and there may be mental operations that are performed too rapidly to allow of their stages being differentiated in introspection. Indeed mediateness of the genuine kind may pass over into immediateness in the spurious sense at various rates: from the slower, involved in reflective processes, to the more rapid, such as may be involved in intuitive induction if that be, as has been conjectured, a case of rapid experimenting in imagination. In any case rapidity is always relative and knows no absolute limit; and rapidity of intake and synthesis is irrelevant to the truth or validity of what is taken in. Consequently it is of no great epistemological import, and is no ground for attributing superior truth-claims to apprehensions which it characterises. If it be said that it is not mere velocity that is concerned but rather absence of all linkage, or mediation such as inferential procedure, one may reply that it does not follow that linkage is absent because it is indiscernible or undiscerned. It is only on the supposition that this is so, however, that the immediacy of intuitive induction can be assigned epistemological significance. Immediacy, in short, is always relative. Even what empirical psychology must take for absolutes, viz. *minima sensibilia*, the specious present, etc., are relative to *homo mensura* and his contingent limitations in respect of *tempo* and discrimination. Similarly 'self-evidence', unless it is an elliptical expression for immediate evidentness to some mind or minds at a certain level of intelligence or stage of mental furnishing, is a non-significant term. We conventionally bestow it upon truths which strike most minds endowed with some indefinite modicum of culture in what is, relatively speaking, a flash. Truths are self-evident to a mathematician which only become evident to me after laborious study and inferential proof; while certain likenesses and equalities that leap to my eyes are invisible to many savages.

I may remark, by the way, that another question may be raised about intuition that is wont to be called immediate, besides those of rapidity of apprehension and absence of logical linkage: that as to whether what passes for pure or *a priori* intuition, within the fields of science and even of mathematics, is really as independent of empirical observation and memory thereof as is wont to be believed, or is actually an instance of experimenting with images or ideas such as presupposes experiments with sensible things. But this topic may be reserved until the connexion of the pure sciences with those more directly concerned with actuality will receive our attention.

It has been necessary to pursue the foregoing inquiries in order to sift the doubt that is commonly entertained as to the capacity of a genetic study of knowledge-processes to throw light on the nature and validity of knowledge. I have tried to show that the objections of an *a priori* kind that are relevant to the genetic method involve dogmatic assumptions for which psychological science knows no ground, while those which possess force or plausibility are only relevant to sensationalism and the naturalist's fallacy, which the truly genetic method eschews. These general objections having been met, I can proceed in the next lecture to set forth what the genetic inquiry into the knowledge-process reveals as to the nature and validity of our presumptive knowledge, or such knowledge of actuality as we have. I will then submit the reasons for believing that all this knowledge consists in a venturesome advance beyond the primary certainties of an individual's experience at a given moment, through postulations and anthropic interpretativeness which can receive pragmatic verification but no logical certification; and consequently that it is probable belief, the probability of which is in the last resort of an alogical kind. And the disclosures thus made by application of analytic and genetic psychology pertain not only to the nature of what we call knowledge but also to its validity. For, firstly, they shew that this knowledge possesses neither of the two kinds of certainty which we are able to distinguish. And, secondly, they indicate the range within which the knowable relevance or validity of concepts obtains; viz. the contexts within which those conceptions originated and to which they are experimentally found to be applicable.

The genetic method, for which I have in this lecture claimed the capacity to reveal what our knowledge actually is, is largely identical with the historical method of which Sidgwick complained as having mischievously invaded every field of thought. From the consideration of it one may naturally pass on to discuss the relation in which history stands to the other departments of knowledge; and this inquiry will also be prosecuted in the near future. There is, indeed, a weightier reason why this topic should be the next to be pursued in an exposition of the philosophy of the sciences. For if the empirical and psychological science of knowledge, in general, stands first in a systematic arrangement of the sciences because by it alone we may learn what knowing really is, history, in the broadest sense of the name, must come second in order of presupposition. Psychology almost wholly constitutes the science of knowing; history constitutes the first science of what is known, in that the subject-matter of all other sciences of the actual takes its rise in the historical.

Lecture III

THE SCIENCES AS HUMAN INTERPRETATIONS OF 'HISTORICAL' DATA

I HAVE submitted in a previous lecture that an inquiry into the nature of knowledge is the first business of a philosophy of the sciences and an indispensable condition of discussing the philosophically significant relations in which the various departments of knowledge stand to one another. I have also pleaded that this inquiry can only be conducted by the analytic and genetic psychology of cognition, and have offered a vindication of the capacity of genetic psychology to throw light on both the nature and the validity of what commonly passes for knowledge. And now the occasion has arrived for setting forth what I take to be the conclusions which this psychological inquiry has established.

It is nowadays generally agreed that all our knowledge takes its rise in sensory perception and could not be had without sensation. One of the few exceptions to this unanimity is the claim that religious, and especially mystical, experience involves an immediate apprehension of real but non-sensory objects; a claim which will be considered later. It is also sometimes asserted that there are thought-given realities; but since the doctrine of innate ideas and beliefs has been abandoned it has been generally allowed that ideation actually presupposes perception. We may come to make and to manipulate concepts, as in mathematics, without having sensible objects, or any impressional core, from which ideas such as that of number have been derived, before the mind; and we may acquire knowledge as to the logical relations between propositions concerning ideas without recurring to sense-perception. But we cannot suppose that general concepts and universals are knowable previously to sensible particulars, now that we possess knowledge about the origination of ideas and the stages of our acquisition of them. Apprehension of the simplest relations, such as likeness and otherness, seems only possible when concrete fundamenta are presented for comparison. In fact, otherness of sensory data is the objective source of all our knowledge of spatial, temporal, and numerical relations, while likeness between some of such data is the precondition of scientific knowledge.

If this be so, there can be no knowledge that is *a priori* in the sense of not being mediated or occasioned, *in the first instance*, by sensory experience, however great be the difference between sensory perception and understanding or reason. This, of course, is not the only meaning that the phrase *a priori* has borne. And it has sometimes been said that in this sense of the phrase there has been no *a priori* philosopher, that even rationalists such as Plato and Spinoza did not ignore the facts of sensory experience but only claimed to have discovered certain *a priori* truths by investigating the contents of such experience. That they must have obtained their highest concepts in this way goes without saying, if my previous assertion be admitted; but it is, nevertheless, a fact that in the exposition of their systems Plato and Spinoza proceeded as if they had not obtained their concepts and *a priori* truths in this way. They expressly taught that knowledge of them is attained otherwise, viz. by a faculty of reason and without the mediation of the baser organon of sense. It may be that the claim to *a priori* knowledge in this sense underlies the prohibition, with which I have previously dealt, of all attempts at a psychologically conducted criticism of knowledge. If so, I shall now have wholly cleared myself of the appearance of sophistry when declaring such procedure possible. However, more important consequences than that follow from this first contribution of psychology to the theory of knowledge, and I will presently point them out in connexion with what I shall call the historical basis of all our knowledge.

Meanwhile, we may observe that in a sensation, by which I mean the actuality describable as the simplest possible kind of percept, we encounter one of the very few instances in which a cognitive act is characterised by that perfect certainty which is wont to be regarded as an essential mark of all that can deserve the name of knowledge, though an individual's momentary sensations are refused the name of knowledge. For the sake of clearness I will now conventionally assign to the words 'certainty' and 'certitude', which are commonly used as synonyms, quite different meanings. By 'certitude' I will refer to the state of an individual's mind that is known as convincedness and is announced when one says 'I am certain that...', and may possibly then proceed to utter a false proposition; while by 'certainty' I will indicate the objective character ascribed to propositions independently of whether this or that person believes them, as is expressed by the words 'it is certain that...'. Now in a judgement of simple perception, such as 'I now see blue', the seeing is not merely believing with the maximum of certitude; it is also for the percipient, but for him alone, a case in which certitude is correlated with certainty, or in which convincedness is objectively compelled and is literally an overwhelming, so that assent or belief has no option. The blue surface is simply given, and at this lowest level of perception illusion is excluded. Illusion or error may enter into the higher stages of what is commonly called perception, but which contains conception also, as when an individual asserts his patch of red to be the setting sun; but into apprehension of the impressional core of individual perception at its lowest level nothing but certainty, alogically but objectively necessitated, can enter. All our presumptive knowledge of the external world is ultimately based on data thrust upon us in individual experience, and all the certainty or approximation to certainty that science can claim is ultimately dependent in part on the self-evidence of sense. Yet this primary certainty of private sensation carries us an extremely little way towards the validity of public science. For, in the first place, the *quale* of one individual's impression, such as blue, is absolutely incommunicable to another individual. And not only are the deliverances of simple perception, in which we encounter our primary certainties, strictly private; they are only true, for the individual, of the here and now. They evidence nothing but themselves. No number of infallible and similar sporadic perceptions of blue can supply their percipient with the certainty of the existence of a blue thing, abiding as such in the intervals between his successive sensations. The certainty of sense-evidence is consequently lacking from every stage of knowledge save our first, which, in itself, is not called knowledge at all, and which is certainly something very different from science.

There are other simple apprehensions, presupposing sensation but otherwise as primordial as it, in which certitude and certainty are concomitant. Some relations between percepts can be read off with immediacy, and these 'objects of higher order' are thrust upon us with a necessitation similar to that which marks sensation. For instance, the subsistence of likeness or difference between colour-percepts is given as truth to the understanding with the same directness and inevitableness with which the being of colours is thrust upon our sense. The relations that are thus apprehended with certainty are few. They are those which furnish what are called the formal categories, which are the only ones needed by the pure sciences of mathematics and logic. Moreover, apprehension of these relations is communicable, whereas the *qualia* of the sensible terms between which the relations subsist is not. We can never be sure that we have precisely the same—i.e. qualitatively identical—colour-sensation as another person when he and we look at the same poppy. But we can communicate to him, even if he be colour-blind, that blood and poppies are similar in colour. Thus the few relations which are apprehended as certain are the basis of all communicability, or of science as distinct from individual perception, as well as being the basis of the pure or abstract sciences. But again, we may observe what a little way the certainty of these relations, when added to the certainties of sense, carries us towards science of the external world; for among them are not to be included the relation of substance and properties, or that of cause and effect, without the invocation of which none of our sciences of actuality would have come into being.

It is otherwise, however, with the pure sciences. In them we have formal truths or thought-knowledge which is as much beyond gainsaying as the sense-knowledge, if I may call it so, of the individual as to the here and now for him alone. And, unlike that sense-knowledge, this thought-knowledge is characterised by universality. It is formal because it deals with forms or ideas from which the sensible matter is eliminated, and with the forms of propositions rather than with the truth of their contents. For this reason it is not necessarily, or in the first instance, knowledge about matters of fact posited for us through sensation, but knowledge about the relations of ideas, mostly posited by us. Every branch of mathematics and every system of geometry posits its own ideal entities by definition, or by what are allowed to pass as definitions. The theorems deduced from these entities together with postulates and axioms follow with logical necessity, and a system of truth, consistent with itself, is the result. If we restrict the name 'knowledge', as I have tacitly been doing, to truth about actuality, cognition of which is mediated through sense-impressions, this formal knowledge should not be called 'valid knowledge' at all, but rather 'consistent thought'. There is no *a priori* necessity that it shall even apply to actuality; and it is for actuality alone to determine whether, and to what extent, some of it does apply. Admired on account of its clarity, universality and logical necessitation, this consistent thought, or truth about ideas, was taken, early in the history of philosophy, to be the paradigm of knowledge and accordingly its method was prescribed as that which philosophy should pursue. But inasmuch as fact-data do not enter into it they cannot be extracted out of it. It is thought, but not knowledge. Or, if we enlarge the denotation of 'knowledge' so as to include it, it is knowledge about the ideal, and not necessarily, or through any virtue of its own, knowledge of actuality in the sense of being applicable to actual things. If a geometry involving space with $\sqrt{-1}$ dimensions be ever constructed, that system of truth will probably not lend itself to advance the study of Nature; but the geometry of three dimensions has proved instrumental to physical science.

Thus the psychological analysis of presumptive knowledge reveals that there are two poles of certainty. There is brute matter of momentary fact for an individual subject, with no thought-connexion and no universality; and there is universal truth which contains no matter of fact. If cognition of the former kind is blind or dumb, that of the latter kind is empty. Neither of them is knowledge or science of the actual. And the whole of the so-called knowledge which we use in the conduct of life, the departments of which we are concerned to relate, is characterised by neither of these kinds of certainty. Our belief in it is a case of certitude, for we are prepared at any hour to act on it; but that is a very different matter from our *credenda* being certain.

Further, the psychology of cognition reveals the several entrances of faith or venture into our presumptive knowing, in virtue of which, as Locke announced, our knowledge of the world is but probable belief. I will briefly recall the chief of them. In the first place, the whole fabric of our knowledge rests on the trustworthiness of memory, in the sense of reminiscence. It is, however, only when memory is of the immediate past, or but emerging out of retention, and the possibility of obliviscence is *nil*, that memory shares the certainty of the sense-impression. As soon as reminiscence has ceased to be impression-sustained retention its objective control may become imaginal instead of impressional, and a rift may occur between certainty and certitude; honest witnesses often confound imaginations and inferences with remembrances. Unless many reminiscences were true there could be no knowledge; but there is no *a priori* principle to determine which memory-judgements are true and which are false. From the outset, knowledge thus involves somewhat of alogical trust, which is justified only by its results. Again, among the relations and the fundamental categories, constitutive of our presumptive knowledge, are some which are not immediately read off from percepts. These are in the first instance read into our data, are supposed by us rather than posited for us, are necessary only in the sense that they are needed, and are justified by their success. It is neither with alogical compulsion nor with logical right that, e.g., the causal relation, when interpreted as efficiency or necessary conditioning, is invoked. Causation, as something more than temporal sequence, is "imperceptible", as Hume observed. Once more, the human belief in Nature's uniformity obviously arose through trustful hope or venture rather than in a direct reading of Nature's secrets, and was acted upon before science and logic were born. And we now know that the unconditional truth of scientific inductions rests upon specific principles which are not *a priori* or demonstrable, and whose probability is not of the mathematical kind, nor—I would maintain—of the logical kind, but essentially alogical and psychological.

Psychological 'knowledge' concerning knowledge, such as I have been reciting, shews several theories of knowledge, current in the past, to be erroneous. For instance, the positivism which declared science to be a manipulation of impressional data by formal logic, and, therefore, to be characterised by both the kinds of certainty with which we are acquainted, and by nothing less certain, is untenable because private impressions are not identical with, or parts of, the public things which science calls phenomena, and because science employs as indispensable for its researches, as distinct from subsequent expositions of its achievements, various concepts over and above those which suffice for logic and mathematics. Again, the rationalism which seriously asserted the deducibility of a science of the actual world from the possible and ideal took the consistency of thought with itself to be the same as the validity of knowledge or relevance of thought to actuality; it identified thought with knowledge—which the psychology of cognition forbids. And, once more, the critical philosophy of Kant, according to which the skeleton of the body of physical science should be knowledge of actuality endowed with the certainty, necessity, and universality which mathematics enjoys, is convicted by psychological science of error as to psychological fact. For the thought-factors which, over and above the formal categories, enter into sciences of the actual, viz. what Kant called dynamic categories, are now known not to be *a priori* either in the sense of innate, or of being of exclusively subjective origination, or of being forms independent of matter, or of being logically derived. They are mediated by the subject in commerce with his body and other objects, anthropic postulations evoked and pragmatically verified by sensory experience but not logically certified, and are—to use Kant's own terms—regulative and not constitutive.

Elimination of these several faulty views as to the nature of our presumptive knowledge brings us to what must be at least an approximately true account of it. For the eliminating-process has consisted in an appeal to facts and has thereby resulted in the establishment of what we call knowledge. I venture to think that the knowledge which I have set forth concerning the nature of our presumptive knowledge is as stably fixed as, say, the chemistry of water. It will probably be as little affected by further advances in relevant science as will that portion of physical science. Pursuit of the *ordo cognoscendi* and use of the analytic and genetic methods of psychology have actually succeeded, it would seem, not only in tracing the origin and developement of our presumptive knowledge, but also in revealing its nature or constitution and in shewing the relation in which its validity stands to the two forthcoming poles of certainty.

I may now enlarge a little upon another outcome of our genetic study: viz. the fact, of which as yet only one or two glimpses have been afforded, that what we bestow the name of knowledge upon consists largely of human interpretation, and would not be forthcoming without the interpretative embellishment with which the human mind invests its data or primary certainties, in order to construct a rational or intelligible world out of them. There is an element of truth in the saying that "all fact is already theory", even when 'fact' includes the relatively crude or undeveloped percept. For the perceptually real generally contains more than temporally present and impressional data, and is thereby given some meaning or reference to what is beyond it. It is intermediate between the simplest actual percept, i.e. the momentary sense-impression, and the idea, because it is subjectively fused, tinged with the incipiently conceptual and enlarged by retention or memory. But interpretation or reading-in enters more manifestly into what is called perception when, as distinct from the reception of discontinuous sensations, an abiding thing is said to be perceived, and qualities, powers, etc. are attributed to it. If one gets the first notion of permanent individuality from the practically continuous presentation of one's own body, and the germinal idea of substantiality from the quality of solidity or resistance to touch, the transition from the barest kind of sensation to perception of things is rendered possible by interpretative supplementation of what sensation dictates, and is dependent on the contingency that our souls are embodied. It is similar with the notion of causal action, which we could not conceivably attribute to things unless we had first 'felt' it in physical efforts. It is from these germinal notions that the explicated and refined categories of substance and cause are derived; and, unless the developement of the refined category out of the crude notion involves at some stage discontinuity or abrupt substitution, for which observation provides no evidence, the postulation and humanisation that are evident enough in the initial notion will persist, in spite of refinement, in the elaborated category. Thus the very 'things' of which the sciences treat are apprehended as such through acts which are to be called cognitive but which are not knowledge-acts, if knowing consists in the certain reading-off of reality. Again, it is hard to believe that our presumptive knowledge of other selves with minds, which underlies all science, is acquired in the first instance otherwise than ejectively or interpretatively in terms of our first crude self-knowledge; for if we directly apprehended other people's thoughts and feelings as we apprehend colours and noises the consequences would be appalling. All these interpretative ventures, or instances of believing when we cannot certainly know, are justified by the success which attends them, and are so overwhelmingly justified as to make us apt to conclude that they are immediate and certain cognitions.

Our categories of substance and cause, like that of end or purpose, have their beginnings in crude individual experience; and that is doubtless why Kant failed in his attempt to deduce them from universal or common experience alone. And they are not of logical origination but emanate from the active side of experience, which is determined by feeling and striving, at least as much as from that of intellection. Experience is possible without them, and they are *a priori* necessities only in the humble sense that we *must* have them *if* we would have scientific experience, such as men perhaps began to seek for practical reasons, and went on to seek in order to satisfy the human appetite for rationality. This sort of experience or knowledge, then, we get by interpretatively assimilating the world to ourselves in order to understand it as human beings can understand. We do not understand the world in the sense of analysing it into pure sense-data and purely logical relations—that, indeed, the world itself forbids—but in the sense of coming, sympathetically and humanly, to some sort of understanding with it; and that essentially consists in finding identities in the world's diversities, permanencies in its flux, and necessary coherence in its contingencies.

Thus, in the mere apprehension of the objects with which the various departments of knowledge deal we find more than pure cognition, immediacies, and necessities. Experience is more comprehensive than cognition, which is but one kind of experiencing; and actuality is richer than thought. The pure intellectualist would fain see, in what we are wont to call knowledge, nothing but the cognitive kind of experiencing. But, as a matter of fact, our knowing, and the thinking which has actually advanced knowledge, are conditioned by interest and action. It is experience as a whole, or the mind as a whole, that enables the intellect in particular, or as its part, to discover truth about the world; and if our knowing were the purely cognitive and logically controlled process that some philosophers would have it be, our sciences would not be entitled to the name of knowledge. We are told that some kinds of knowledge are conditioned by moral sensitiveness or by sympathy and love; and we shall be the more inclined to respect that statement if we have become convinced that alogical factors and interpretativeness enter into the foundations of all that passes for knowledge and into processes, such as the inductive method, by which raw data are elaborated into systematic science. Pragmatically verified interpretation is the nearest approach there is in our actual knowledge to that logically certified apprehension which pure intellectualism inserts into its definition of ideal but non-existent knowledge. It follows that our knowing involves somewhat of humanisation. In other words, between science and the world stands human nature. Our knowing is specifically human, and its commonness —often called universality—is still commonness to men. The world is knowable only to an extent determined by the range of human faculties and only in a manner that is conditioned by their distinctively human nature. Man is thus the measure, not of all things, but of all things as they are known by him. We have already seen that sensory perception, in which knowing takes its rise, has its nature determined by human organs and that some of the categories that are indispensable tools used in all knowing of actuality are dependent on our embodiment and on the interaction between subject and objects, as well as on brute fact rather than on logical conditions. And now we may further observe that intellection, at its higher stages, where the limitations of human sensibility, the prescriptions of human *tempo*, etc., are of less account, does not effect a dehumanisation of the human mind, nor a transcending of distinctively human modes of assimilation. For instance, if motion be the concept which 'lies nearest' to our understanding, and on that account has been adopted as a fundamental *explicans* in physical science, it nevertheless involves the anthropically derived notion of permanent substance; if rigidly mechanical explanation be the quixotic enterprise of physics, such explanation is not compelled by facts but is a quest for a human *desideratum*; and if simplicity be a canon of explanation in physical science, the impulsion towards simplification of laws by economy of concepts and paucity of independent principles comes not from Nature but from human nature. Nor do the conceptions employed in theoretical physics, or in mathematics, carry us beyond our human conditionings; they are, more or less, arbitrary conventions posited *ad hoc*, and minister to the making of ourselves intellectually at home in our world, just as some of the beliefs of popular religion serve to make people emotionally at home in it. All this is tacitly repudiated, indeed, by what has been called "the great tradition" in philosophy, which seems to imply that no knowledge of reality is possible unless the dehumanisation of knowing, or rather of opining, is already accomplished. So the great tradition has taught us not to regard concepts as what the empirical science of knowing tells us they are, viz. as our mental abstractions from particulars which *happen* to recur or to resemble one another, but as transcendent, or prior to cognition, and as constitutive of ultimate reality. Conceptions derived from, relating to, and valid of, things were hypostatised into things or existents, and taken for the ultimately real things. Symbols, by means of which Nature can be interpreted or made assimilable by our minds, came to be regarded as the constituents of the intelligible and real world which is seen darkly through the temporal and phenomenal. This tradition was established before the nature of our knowing was empirically studied or adequately investigated; and reflection on our presumptive knowledge, as we find it, shews that knowledge, as conceived in that tradition, is but an ideal of what a pre-scientific but eminently rational age would fain have knowledge be. Such reflection also suggests that the phrase 'universal experience' is a dogmatic hyperbole, and that we should do well to replace it by the phrase 'common experience', which begs no question. Even if there were experience actually and necessarily possessed by all mankind it would not follow that its truths were trans-human, or independent of all mankind; for the over-individual is not necessarily over-social or absolute. But, as we have seen, our common knowledge of the actual world is not characterised by certainty and necessity; and it is not transcendent of our humanity or even of elements in it other than the supposed infallible *lumen naturale* and the alleged capacity to see "in God" or *sub specie aeternitatis*. It is not even wholly and purely cognition. It is rather the output of the whole man, or of the mind as feeling, desiring and striving, as well as being aware. I do not mean merely that man has *rapport* with the world other than that of the purely cognitive kind, but that other kinds of *rapport* than the purely cognitive are constitutive of the composite thing which we call our knowledge. There is no science of actual things that does not involve anthropic analogising and faith—which is at least as much conation as cognition. The probability attributed to the ultimate postulates on which induction rests seems to be entirely alogical, a matter of the psychologically inevitable rather than of the logically necessary, or a case of the hope that springs perennial in human breasts. Consequently the basis of the whole body of what we account to be knowledge is in part non-cognitive. In making this assertion I am not wantonly wallowing in irrationalism; I am stating what is thrust upon us by hard facts and what I am convinced the logician will be compelled to admit without reservation. Nor should the assertion in question unduly shock our intellectual pride. The anthropic nature of our knowledge and its saturation with the non-cognitive do not debar it from claiming to be a version of the truth about reality. For if the correspondence between judgement and reality, in which truth is generally held to consist, be relevance rather than identity or copying, the phenomenality, relativity, and specifically human assimilativeness, intrinsic to our knowledge, need not vitiate its truthfulness any more than they depreciate its practical value. But perhaps these words of consolation will sound oracular until their import shall have received further explication.

Having now completed the exposition of what our knowing is, I will enter on the discussion of what is known, and of the manipulation of our knowledge-data which issues in the products called departments of knowledge or sciences.

To begin at the beginning, we may recall that the impressional core of our sense-data is the foundation of all objectivity, or of knowledge of what is other than ourselves, and, in the order of knowing, is our primary contact with what the metaphysician would call ultimate reality. It is our first received manifestation of reality to us, its only direct utterance, or, to change the metaphor, the one and only vision that is vouchsafed to us as beholders before we come to see what we think must be there. If ever it be possible to acquire metaphysical truth, or to advance to a purer and clearer, a less phenomenally coloured and a more systematic, vision of ultimate reality, it can only be by reflecting upon, and manipulating and improving, what I may call the chromatic vision mediated through sensory perception; just as in physical science we correct and extend our sensory observation by means of instruments and calculations. It is the nature of our sense-data that dictates the subject-matter of all our concrete sciences, and it is their relations in respect of likeness, difference, plurality, recurrence, etc. that provides the basis of our more abstract sciences. It is the impressional, again, as distinct from the imaginal and the ideal which are derived from it, that supplies the only direct criterion and certificate of actuality. And, as psychologically ultimate and inexplicable, or as simply brutal, alogically posited, and thrust upon us whether we will or no, this perceptual factor rules out the possibility of our knowledge of the actual world ever being made wholly rational, in the severer sense of that word, or being reduced to a pure science. For a pure science deals only with relations between ideas, or between objects from which the perceptual element has been eliminated, as negligible *ad hoc*.

It is a far cry from what is apprehended by us individually in simple perception to cognition of a common world of things, or to experience as organised by common sense. In a subsequent lecture I will describe how transition may be effected from the one to the other by means of intersubjective intercourse, concepts, and postulatory categories, and I will submit that the common-sense realism, according to which physical objects are perceived agenetically, or with the same immediacy as characterises the sensation of a colour, is incompatible with the facts presented by various abnormal types of sensory experience. But, leaving this gap to be bridged in the future, I would now pass without delay from actuality in its primary sense of the perceptual in individual experience to what we commonly mean by actuality, viz. the world composed of the abiding and interacting conceptual things of which common sense and science speak. The sole claim of this order of phenomenal things to actuality, which marks it off from the realm of non-actual entities such as those with which pure mathematics deals (and which are likewise conceptual), consists in the fact that our constructed concepts of things and events, with all their suppositional elements, are saved from being empty ideas by the core or nucleus of the perceptual which we each enclose in them. Using the term 'historical' in the wider sense which is now current, to denote the actual and changing, the concrete and particular, the qualitative and maybe the unique, as contrasted with the universal, the abstract, the timeless, or the quantitative, one may observe that the historical sciences, or the historical portions of each science, are those which set the questions of philosophy. A philosophy which abstracts altogether from the historical is a facile pursuit which ignores the essential character of the world with which it should be concerned, and evades the peculiar problems which our particular world presents in the bustle and confusion of historical happenings. For the reason that the historical is our first collective knowledge-datum from which the sciences and philosophy must set forth, history, in the wider sense, is the department of knowledge which occupies the next place, after the science of knowledge itself, in the order in which the sciences are to be arranged by a philosophy of them such as would build on solid foundations.

In other words, history takes the first place in the systematic order of the departments of knowledge dealing with what is known, as distinct from our knowing. For its matter is determinative of all the other sciences of actuality, and they are outgrowths from it; also it prescribes the method which philosophy, as distinct from a pure science, must follow.

This is a somewhat unorthodox view. For philosophers have evinced a general tendency to disparage the historical. They have done so on the grounds that it is a medley of contingent and particular facts, known by perception and memory rather than by reason, and that it is not amenable, as such, to science which deals with the common and repeatable, or to philosophy which would connect ideas and propositions by ratiocination or according to logical implication. But this disparagement seems often to have issued in a setting up of conceits in Nature's stead, or of tidy but non-actual worlds in place of this world of rough and tumble. So far from being negligible, the historical possesses import for both science and philosophy in several respects that are relevant to a discussion of the relations between the various departments of knowledge and thought. We must recognise at the outset that an alogical factor enters into the foundations of all our knowledge. The *posita*, of which I have already spoken, are prior to logic and determinative of all possibility. No reason, let alone a logical or an *a priori* reason, can be assigned for their being what they are, or indeed for their being at all. Their particular determinateness is inexplicable. No accumulation of universals or empty concepts will produce one qualitative particular. Neither from Plato's supersensible world nor from Spinoza's one substance, any more than from the self-identical reality of Parmenides, is there a possible derivation of the historical, the concrete, the diverse, and the particular. Qualitative essence cannot be extracted out of idea, form, or quantity, because it is not contained therein. The alogical factor in what is primarily known is thus irreducible and non-deducible. It is also non-rational in that its order and connexions are not identical with those of thought. The world of mechanical science is reversible, while the historical is essentially irreversible. Brute facts are not predictable in the first instance; it is only after acquaintance with some of them and with their experimentally discovered relations that science can come to predict others. And the relation of fact to fact is not that of logical implication; the *rapport* between cause and effect, whatever it be, is not the relation of ground and consequent. Thus, there never can be a science or a philosophy of the world that is rational without remainder; for the world owes its very essence largely to its irrational or surd element. Nor can historical developement, which is temporal and irreversible, be reduced to a dialectical developement which is timeless.

One might at this point pursue several courses of argument which lead to similar results: e.g. that the notion of possibility presupposes some given actuality; that laws are not a self-subsistent *prius* determinative of what this or that thing must be, in order to be at all, but simply statements of how things have behaved, making their laws by their behaviour, and obeying them by being regular in their behaviour; or that change and becoming, involved in history, could not even appear to be if the real, of which they are the appearance, were immutable and *ipso facto* not productive of the appearance. These conclusions contradict the doctrines of such philosophy as slights the historical; but the slight is motivated by a desire to find the world rational or logical all through, in spite of appearances—or rather facts—to the contrary.

At the same time there is another side to the alogicality of our historical world. In addition to its capacity to wreck all attempts at complete rationalisation of the world such as would consist in reducing its diversity to identity, it is the source of much that a world-view cannot overlook. Though it is not rationally comprehensible it is not irrational in the senses that it is nonsensical or meaningless and that nothing can be made of it. It is matter of knowledge, or of what we commonly account knowledge, if not scientifically knowable according to a common meaning of 'science'. It can be understood in the sense that by sympathetic *rapport* we can come to an understanding with it. It can perhaps be explained teleologically if not logically. But what is most important about it is that it is the carrier of value, and, in consequence, its entrance as a factor into this world of ours redeems an otherwise rational world from being meaningless and unreasonable. A world of static universals, immutable reals, homogeneous or identical individuals, might be logically intelligible and mathematically calculable, but it could have no meaning, or serve no purpose, and it could realise no good. Fortunately our world of brute happenings has an interest for others than contemplators of empty eternal verities: for beings that can appreciate values, pursue ends, and cherish affections and hopes. And the fact that the happenings and conjoinings in the historical world of Nature and man are not logically compacted and are not scientifically knowable, does not involve that the world is hopelessly baffling and stultifying to beings which possess other faculties and capacities than that of pure and formal thought. We have already seen that our knowledge of the world and man is not composed entirely of pure and logical thought working on necessitated data that are the same for all; and we can now see that science such as seeks only the common strands and quantitative relations, however great be its comprehension, is inadequate to cope with the fullness and richness of actuality; while a philosophy which would strain out the historical as irrelevant to higher knowledge leaves the distinctive essence of our world, and the temporal order of becoming, entirely out of account. On the other hand, it is the historical, with the manifoldness and variety of which our knowledges try to cope as best they can, that is the source of all law and quantitative relations, and of all value and significance for personalities.

In fact, all sciences begin with the historical, and with empiric facts as they are constituted at the level of philosophical organisation denoted by common sense. This is as true of modern physics as of the logic and the science of Aristotle. Natural history precedes natural science; and some departments of science, such as geology, astronomy, and evolutionary biology, are concerned with the past history of Nature, tracing the changes through which larger or smaller parts of it have successively passed, and its linear series of causal connexions between individual events, just as the historian is concerned with past events and causal connexions in the sphere of human life. We are wont to place such departments of natural knowledge lower in the hierarchy of the sciences than those, such as physics and mechanics, which treat of the common and the repeatable, the general and the abstract, and which can express their conclusions in the form of laws and equations. In doing so, however, we should bear in mind that we are arbitrarily adopting one criterion as to rank while another is equally open to us. If nearness to concrete actuality and retention of its richness were our standard, rather than the exactness and economical symbolism which are only attainable when we eliminate all aspects save one, the more abstract sciences would take the lower rank. Comparison, such as is expressed in the terms 'higher' and 'lower', or in the words 'more essential' and 'less essential', implies relativity to some particular interest, not absoluteness, or independence of all interests and ends. And as to the essentialness of the historical for philosophy there are two points to be considered. Firstly, the outcome of the historical or genetic method of inquiry may be fraught with great philosophical significance; a question presently to be discussed. And secondly, we have no grounds for identifying degrees of abstractness with degrees of approach to ultimate reality. This time-honoured prejudice still flourishes, but it seems to be as vicious as it is gratuitous. In the ontal there may well be more of structural detail than is manifested in the phenomenal; but there can hardly be less, unless men, like God, are creators *ex nihilo*. We cannot make phenomenality without provocation and co-operation from the ontal, since an appearance must be an appearance of something as well as an appearance to someone; nor unless there be some correspondence between the structure of the ontal and the structure of the phenomenal: thus much of metaphysics has always been tacitly presupposed by science, and is essential to it. Historical knowledge, therefore, for all its chromatism, gives us a far more adequate mirroring of reality than do the abstract schemes of the 'higher' sciences, which ignore so much of what is to be reckoned with by philosophy.

The word 'history', which I have been using in a broad sense, is generally used in a narrower sense, and is restricted to the sphere of human life. In human history the beings concerned are agents, as contrasted with the static universals, inert particles, timeless laws, etc. which figure in the more abstract and non-historical sciences. They are also individuals, not in the sense of qualitatively identical though numerically diverse instances, as are the physicist's atoms, but as possessing idiosyncrasies, different interests, appreciations, and so forth. And they exist at definite times and in certain places. Also the groups in which these individuals are included, such as nations and parties, are organised wholes and not logical classes in the concepts of which the content, or the intension, varies inversely with the extension. It is the deeds and utterances of these agents that constitute the facts of human history. Such facts are due to the self-determination of their agents; and they are particular, concrete, and unique or once-occurring events, as contrasted with the abstract and repeatable features of events which are contemplated in mechanistic physics. It is conceivable that all happenings throughout the physical world are historical in this sense, and that regularity and progressive developement in Nature have been brought about in a manner similar, in some respects, to that in which they have been brought about in the case of human individuals and societies. If so there will not be complete disparity between men and what we call matter. One may observe that, quite apart from such metaphysical speculation as would interpret material phenomena in terms of spiritual monads, there is a growing tendency among theoretical physicists to historicise the classical mechanics. The ultimate 'continuants' in the physical world are being conceived as individuals each with its own biography, its mass being as dependent on the present phase of its adventure as is a man's temper on the latest vicissitude in his affairs, its quality and behaviour being determined by its own *tempo*, etc., and the geometry which used to be taken as a set stage for the play of atoms being regarded as the outcome of their movements. From many quarters we are being told that the old static systems in terms of which Nature was scientifically explained are incredible, and are but attempts to describe a historical world as if it were unhistorical. It is not necessary now to pursue such metaphysical speculation;

but we may observe the fact that the events into which we differentiate the course of the world are, as we most directly apprehend them, qualitative and irreversible occurrences, and are so far historical in the sense in which I have been using that word. The historical, on which all our departments of knowledge are founded, fills up the many gaps left by the abstract sciences, or is rather the concrete totality of which the abstract sciences study only certain aspects. Philosophy, which would see things whole and as a whole, cannot take its departure from sciences which owe their very existence to leaving out much of the stuff of knowledge because it is not scientifically manageable by them, or is irrelevant to their business. What such sciences leave out is still there. And if their selected data yield the conclusion that the physical world is ordered by number and law, or evinces logical rationality, the data which these sciences discard and relegate to history and other departments of study are not altogether incapable of being interpreted or explained in another manner, and of being systematised so as to reveal a genetic order, suggesting that the world's rationality is overlaid with reasonableness or meaning. The historical, in other words, may be a realisation of what are called the eternal values, and certainly is a realisation of values. On these accounts, and on others previously mentioned, history is not a department of knowledge that is merely ancillary to philosophy, as philosophy was once accounted to be ancillary to ecclesiastical theology. It is rather the source and the determinant of philosophy. Philosophy has, indeed, sometimes asserted that meaning, in the sense of purposiveness and realisation of value, cannot be found in the temporal order, and therefore, for the satisfaction of human hopes and aspirations, needs to be postulated in a transcendent or supra-temporal order; but philosophy then seems to me to have uttered the reverse of the truth, and to have done so through reverence for abstract science coupled with neglect of the rock whence such science was hewn. Attention to the historical and to its message, or to the primary and humanly unsophisticated deliverances of the ontal, which cannot be identified with timeless abstractions, is corrective of this partial conclusion from partial data. Such I take to be the significance of history, in its broadest sense, for a philosophy of the world and man. Hence I would assign to history, on the objective side of knowledge, a position similar to that which I have assigned to the psychology of cognition on the predominantly subjective side of knowledge, viz. that of a first propædeutic to philosophy and a first science in a systematic ordering of our departments of knowledge. The exposition of the reasons for this estimation of the historical, and the procedure by which the historical may be investigated so as to give an answer to the question whether we may reasonably regard the universe as embodying a purpose—which is ultimately a theological issue—constitute what I would understand by the phrase 'the philosophy of history', in which, of course, human history is to be included. I need hardly say that such philosophy of history is something different from various forthcoming philosophies of history, which have rather been attempts to fit historical facts into a scheme of dialectical metaphysics, or into a system of dogmatic theology, constructed in aloofness from what the historical facts tell us of themselves.

From history, in the generalised sense in which it includes natural history and actual happenings of every kind, I will pass, in my next lecture, to history in the more usual sense of knowledge concerning human affairs. In this narrower field the word 'history' is a name for two different things. By a man's history, e.g., we sometimes mean the sum of his successive acts, utterances, etc., or his *βιός*, and sometimes we mean a narrative of his life-activities, or his biography. History, in the latter of these senses, has been called historiography. It is one of our departments of knowledge; and as it has recently been a vexed question among historians whether it should be called a science, its relation to other knowledges calls for some discussion.

Lecture IV

THE RELATIONS OF HISTORY AND DOGMATIC THEOLOGY TO EACH OTHER AND TO THE SCIENCES

In my last lecture I was discussing our empirical knowledge of the historical, in the wide sense in which history embraces natural history and all kinds of actual happenings, and the import of that knowledge for a philosophy of the sciences. I will now pass on to the narrower field of human history and consider the question whether history, in its commoner and more restricted sense, is to be regarded as a science. It is better to put this question in another form, and to ask in what relation history stands to other departments of knowledge such as those which are wont to be described as sciences. For whether history is a science depends on how we agree conventionally to define the meaning of 'science', which, as ordinarily used, is an elastic term. If certain large tracts of astronomy, geology and biology are entitled to the name of 'science' which is usually allowed to them, it will seem hard to refuse the name to human history. For geology, etc. are concerned, like history, with establishing particular and once-occurring events, and use essentially the same method as does the historian, viz. that which may be described as tracing a past event by its effects. I should perhaps observe, at this point, that by 'history' I mean statements concerning the historical as I have previously defined it. For there is what has been called history, and has been allowed to be science in the stricter sense, which is not history in the sense in which I have been using the word: e.g. Comte's social mechanics, his 'history' without names, events or dates, and the generalisations of anthropology and folk-psychology. Dismissing these more abstract and general topics as not belonging to history properly so called, I may further indicate the nature of the claim of the latter sphere of study to be called a knowledge, if not a science, by remarking that processes, of which events are individual stages, can be investigated by a method which is apparently capable of yielding truth or knowledge. And this method is a part of the complex thing called the scientific method. It traces causal connexions between particular events; and we can scarcely doubt that it yields some knowledge of facts. Further, history is not a mere inventory of items of fact. Like other sciences it systematises its facts, though not with the same type of system. It arranges them not only in chronological order but also according to causal connexions. And such connexions serve for a genetic treatment of posterior facts, institutions, etc., which endows them with an interest or a significance that would be lacking if their causation and gradual developement remained unknown. It is a relatively unimportant matter whether or not we restrict the name of science to such departments of knowledge as deal only with the common and repeatable features of the historical, and accordingly exclude history from the sciences, so long as we recognise that history is systematised knowledge of facts, revealing more about them than their chronological order. This further truth is the knowledge which we call genetic. And, in virtue of its being acquired by historical investigation, the phrases 'historical method' and 'genetic method' have become generally synonymous. They denote a mode of inquiry which is relatively new and which has profoundly influenced many fields of thought.

Roughly speaking, history passed a century or so ago out of the stage of critical sifting of traditions and documents, of piecing together the facts so established, and elucidating them didactically. It also shifted its main interest from the relatively narrow field that had previously occupied it. It began to ponder over events of many sorts as stages in a process of developement, and to bestow more attention on the mental or spiritual principles which controlled or shaped events and movements. It thus revealed the continuity of activities with activities and of age with age. Concurrently, Lyell and Darwin studied wide tracts of Nature from a similar point of view. Thus arose a genetic mode of thought, sharply distinguishing the mentality of the nineteenth century from that of the eighteenth and earlier centuries, whose thought was largely dominated by static ideas, and whose outlook on products of developement was unhistorical, in the sense of agenetic. Many studies were thus revolutionised, being approached from a new point of view and pursued by means of a new method. Indeed the historical movement has exerted a determinative influence upon thought and investigation comparable in importance with that of Newtonian science and the Copernican revolution. It has done so by furnishing us with two potent ideas, both of which are involved in the current notion of evolution. The one of these is the idea of continuity, applying to changes in Nature, to all mankind, and to all ages including the prehistoric; and the other is that of emergent novelty, epigenesis, or growth determined by influences from without and not solely by self-unfolding of the preformed. The historical, as interpreted by these two ideas which itself has thrust upon us, largely redeems itself from the apparent promiscuity and mere contingency, on the strength of which science and philosophy had somewhat superciliously looked down upon it. There is a method in its madness, or a system in its hurly-burly, revealed in spite of the fact that it cannot be formulated in abstract laws and in terms of the universal and the logically necessary. And it is a question, though a vexed question, whether or not the guiding ideas supplied by history are not as essential to philosophical knowledge of the world and man as any of those, such as law and uniformity, which are furnished by the sciences of the general.

Some of us would boldly subscribe to Aristotle's doctrine that we must know the generative principle, if we would understand the essence, of a thing; or would at least maintain that we *may be* erroneously conceiving the nature of anything until we know its history, or how it came to be what it is. Certainly we can point to numerous instances in which erroneous statements as to the natures of things have been revised or corrected in the light of knowledge as to their genesis: e.g. the notions of laws of Nature, of natural religion, of knowledge, reason, conscience, and certain human institutions, which obtained in the eighteenth century. I have previously represented that the dynamic categories, as Kant called them, have their natural history, and that indifference to that fact led him to mistake their nature and to 'derive' them fancifully. One might call some of Kant's categories habits of thinking that have proved useful *ad hoc*. And similar representations might be made as to the forms of intuition, or our notions of space and time. Historical inquiry shews us that such instruments of thought need not be fashioned with immediacy, or be apprehended as ready-made, in order to be forthcoming, nor need be universal, necessary, and nonsupersedible in order to apply when and where they do. And one may go further and say that our rough notion of a thing, which we make static for logical convenience, is then only a tool for relatively rough purposes; for Nature is ever in a flux, and what we call 'a thing' is but a thinner or a thicker cross-section of a process or of a life-history.

An oak, e.g., is any or every stage of transition from an acorn to a huge dead tree. Logic tells us that a thing must either be or not be; but if, as science encourages us to believe, everything is in a state of continuous becoming, there are actually no things such as those to which logic refers. So the outcome of a career up to date, taken to be a finished product if not an entity like Melchisedek without descent or parentage, does not suffice to give an exhaustive idea of that career as a whole. Static ideas are the only counters which logical thought can use; but if the historical or evolutionary view of Nature be accepted all logical thought, as applied to Nature, bespeaks makeshift, approximateness, and inadequacy. The logically perfect idea is rigid and clear-cut as a crystal; the scientifically perfect idea, though it could hardly be used logically even if it were forthcoming, would rather be colloidal: prospective as to potentialities to be actualised in the future and retrospective as to stages that previously supervened. Thus statements as to the nature of things, once things are no longer conceived statically, are partly statements as to origin and developement, unless they confess to being largely abstract or artificial.

Such considerations as to the relation of history to logic, science, and philosophy, are suggested by the self-manifestation of the historical in respect of its continuity and its fluidity. Others are suggested by its self-manifestation in respect of epigenesis, or what is now often called emergence. With this feature of Nature, again, rigidly mechanical notions, which in turn are based on ideas of the static and identical, cannot cope. A change in the physical world bespeaks previous changes; but the antecedent changes are not, in the case of organic phenomena at least, a sufficient datum for prediction of what will follow. There is a margin of the emergent and the contingent, or the non-mechanically necessitated, which at the same time is not ascribable to lack of conditioning, or to what is called chance. Not to dwell now on this aspect of Nature and its import for science, of which recent literature has been making much, let it suffice to observe that history, in bringing it into prominence, makes a demand on philosophy for a profounder treatment of causation than physical science, which never penetrates beneath the phenomenal aspect of causality, can supply. But I have reverted to history in its broader sense, and must now return to human history.

In the light of what has previously been said, human history may now be described as the investigation and exposition of the causal connexion of facts concerning the developement of men as social beings. We may study its nature and method by examining piecemeal the grounds on which some recent historians have refused it the name of a science.

The first stage of a historical investigation consists in proving the actuality of alleged past events. The severer critics emphasise that the historian's primary data are not first-hand observations, like the data of the physicist. At the outset the historian is not confronted with what took place, but only with what *A*, *B* and *C* are said by *X*, *Y* and *Z* to have done. So, it is urged, historical truth is at best correspondence with documents, not with facts. The historian's handling of original records gives a false sense of security, as if the assertions contained in original records necessarily were accurate accounts of pure facts and, when critically sifted, were free of all elements such as may be called products of art. It is implied that it is impossible to establish bygone events, let alone to "describe them as they really happened". Moreover, the testimony to alleged facts in the remoter past is said to be generally such as would not now be accepted in any court of justice, and it is pointed out that our documents often give us only fragmentary allusions, which are fortuitous survivals, and that this imperfection of the record cannot be repaired. On these grounds the very foundations of human history are declared to be unsafe.

If this criticism only amounts to asserting that history is not a body of sensorily and logically certifiable truth, it is doubtless irrefutable. But, if it is meant to imply that a sharp line can be drawn between history and all the presumptive knowledge that passes for science, it is to be accounted hypercriticism. For, as we have seen, all our presumptive knowledge of actuality falls between the two poles of certainty. Even the physicist has to rely on the work of his predecessors, and therefore on history and testimony. And it does not follow from some testimony being false that none is true. An individual's memory is fallible; but unless some reminiscences were true and capable of being used to correct others there could be no knowledge. And it is the same with what may figuratively be called social memory, or history. It is not logical certification but pragmatic verification that is the proof of science, and not certainty but probability that is the guide of life. To adopt the former criteria of knowledge in order to rule out history, and the latter criteria in order to let science pass muster, is plainly favouritism. Recognising, then, that all knowledge rests on probable belief, history is only to be disparaged in comparison with knowledge of Nature either on the score that it uses an inferior standard of verification or else on the ground that it can establish no matter of fact with probability sufficient to warrant reasonable confidence.

Leaving methods of verification aside for the present, we may pursue further the question of the primary data of history and their relation to fact. The criticism which I cited just now is one-sided in that it ignores two classes of data that are generally forthcoming for the historian of the distant past. Besides second-hand evidence or tradition about past deeds, such as sagas, chronicles, memoirs, inscriptions, and pictures, there are first-hand evidences also, viz. survivals comparable with the geologist's fossils, such as tombs and bodies, arms, monuments, bills, and letters. Certainly there is always the possibility that these have been faked, or there is the principle of the 'plurality of causes' to be reckoned with; but this again applies to all reasoning from the present to the past. But, further, there are also the persisting and observable effects of past events, paralleled in geology by rock-scorings and perched boulders, from which past events can be more or less accurately inferred. Actuality that once was made in then-present experience was made for ever, and neither lapse of time nor human doubt can unmake it. If past events are dead, they yet speak and their works "do follow them". If they *are* not, they *are* yet actual.

They bear fruit, and by their fruits they may be more or less known. Indeed large classes of facts constituting what we do not shrink from calling science are only inferred from their observable effects. No event in the world's course is an ephemeral happening and nothing more. It is also a cause occasioning a chain of effects. And to prove the past occurrence of an event from its effects is counted a fruitful scientific method. It is not necessarily a less scientific method when employed in the sphere of human history than when used to track the past changes of the physical world. Indeed no one doubts that the bulk of the facts asserted by careful historians actually happened, whatever suspicion may attach to affirmations by bygone writers as to other alleged occurrences. History is a department of knowledge though it is not an experimental science.

The second stage of history, in so far as it is investigation rather than constructive art, consists in the critical sifting of its primary data. The historian is not wholly at the mercy of his documents. Though he is precluded from perceptual verification of alleged events such as is generally open to the scientific experimenter, he can apply certain tests to his records of them. He can ask who his informants were, when and where they wrote, and whether they were dependent on others. He can also employ the internal, or higher, criticism and inquire into the individuality and the credibility of the witnesses, their interests, possible motives, 'tendency', etc., and so may find means to discount or to eliminate the artistic and interpretative colouring which is apt to enter into narratives.

Thus far I have been mainly dealing with the relation of the historian's data to facts or events as separate and particular items of knowledge. But the results of historical inquiry are not merely the isolated particulars with which history is accused of being concerned by those who would disallow its claim to be in any sense a science. History is not "all facts and no factors". It is not a linear series of disconnected events, nor comparable to an old almanac. Certainly, as we have seen before, it is not concerned with common or repeatable aspects but with the concrete deeds of individual agents: it does not deal with the general or universal, as such. Hence Aristotle deemed history to be less like a science than is poetry; for dramatic poetry tells what a typical person would do in a generalised situation, and thus yields knowledge that has a kind of universality. Nevertheless it is precisely the continuity of events and their relation to a context of developement that historical inquiry has of late been concerned to emphasise. It does not thereby render itself a science or a systematisation of facts in the same sense that, e.g., chemistry is a science, but it none the less makes itself a body of connected facts, and its connecting thread has been approved as scientific by its incorporation into evolutionary sciences. One recent critic has described history as a catalogue of facts pleasantly rehearsed, and has said that it cannot become a science until it is reduced at least to biology, and its facts can be resumed in scientific formulæ or in terms of hunger and sex. This view of what history should be, however, leaves out of account all the play of mind, reason, and invention, in virtue of which human history is a unique field of study. It thereby offers an instance of the naturalist's fallacy.

As the disconnectedness and particularity of historical facts has thus been somewhat misconceived or exaggerated by critics of history, so also has the contingency of these facts. This contingency, which really consists in the absence of that nomic necessitation which mechanistic physics postulates and, within limits, 'verifies', has often been spoken of as if human actions were merely discharges of caprice, or bolts from the blue. But the alogical element which distinguishes the historical from the scientific is not "against law" but rather without law, in the rigid, false, and comfortable sense in which 'law' was used in the eighteenth century. And with all deference to science, there is knowledge, if no strictly scientific knowledge, where laws are not traceable. Laws are tools for other purposes than that which history has in view. The principle that every effect has a cause may none the less hold in a field of facts within which the other causal principle, that like causes produce like effects, with which the more abstract sciences chiefly work, is scarcely applicable. Moreover the historian, as much as the naturalist, makes use of the principle of uniformity, and also of knowledge as to what is common to human minds, in respect of motivations, etc., over and above knowledge or supposition as to the mental determinants of specific and once-occurring events. However, to keep to the present point, the contingency of the historical is not indeterminateness but self-determination. And it is ultimately from the contingent facts of history, in the broader sense of 'history' that all our 'necessary' or supposedly 'universal' laws as to actualities are distilled. "History", said Lessing, "is one thing, philosophy another. This is the wide, ugly ditch which I cannot pass, often and earnestly as I have attempted the leap." But the ugly ditch would not have presented itself to the eye of Lessing's mind had not his rationalistic predecessors laboriously dug it. That there are eternal truths that do not depend on historical facts, in the sense that they could not be true if the historical events had not happened, may be allowed; but that any such truths pertaining to the actual world are knowable without observation of historical happenings within that world is an assumption to which Lessing was not entitled.

At the third distinguishable stage of the historian's task, viz. the selection, synthesis and interpretation of his established and sifted data, historical research gives place to historiography. An inquiry into the nature of this science, or art, or branch of philosophy, as it has been diversely described, naturally begins with its initial process of selection. There are doubtless questions which history is precluded from answering by the paucity of forthcoming data. But the nearer the historian comes down to his own time the more unmanageably copious becomes his material. Selection is unavoidable. It always was so, indeed, in the ages in which historians were few. Out of the manifold of contemporary particulars even the ancient chronicler could but record a few. And selection is always and necessarily *ad hoc*, or with reference to some subjective interest and some standard of value or significance. Standards, however, change with the enlargement of life: events, etc., which in an earlier generation were deemed most significant and worthy of being handed down to posterity may now be accounted less significant than others. Thus, historians used to bestow attention chiefly on wars and princes, whereas to-day we are more deeply interested in institutions, social progress, acquisition of freedom, and developement of ideas—so that Maitland could affirm history to be not so much of what men have said and done as of what they have thought. Thus even into the selection of data there enters an element of valuation and subjective interest, which varies from historian to historian, country to country, and age to age. Consequently, it is represented, history is precluded from being a science and from enjoying the disinterestedness of science.

I have previously remarked that science's disinterestedness consists partly in uninterestedness in certain interests and partly in pursuing an interest of its own. It is therefore not simply in involving an interest that history differs from the physical sciences. Nor is it in the fact that history is selective; for science is nothing if not selective. The difference rather lies in the fact that whereas scientific investigators, roughly speaking, share one main interest in common, which remains the same from generation to generation, historical writers, of different ages and even of the same period, evince some diversity of interests determinative of their selections and manipulations of their data. Too much, however, can be made of this diversity, actual as it is. It is not unlimited; and selectiveness is not necessarily arbitrary. Nor is it in the sphere of history alone that selectiveness can be abused. Historiography is dependent on, and helped by, all kinds of sciences which exercise some control over its intrinsic capacity for waywardness. And there is, after all, some 'objectivity' in our criteria of the relevance and significance of events. Human belief on such matters may be fluid; but the advance from one belief or valuation to another has not been altogether random substitution. The change has not consisted so much in reversal or destruction as in absorption and gradual developement. The criteria of selection are not exhausted when we eliminate such selective principles as may be dictated by emotionalism, romanticism, or patriotism. It is an axiom, or postulate, of history, as well as of other sciences, that there is a fundamental identity, community, and continuity in human nature, in whatever respects individuals may differ. Objections urged against the claim of history to be a department of knowledge, on the score that history is selective of its data, incur the danger of proving too much.

As for the element in historiography that consists in giving a connexion and interpretation of historical facts as they present themselves to a given age or a particular writer, it would seem that its hypothetical and purely imaginative ingredients have often been unduly magnified. This is the case both with writers who would cast history out of the list of sciences and with writers who maintain that history is a science—despite the fact that historiography cannot be impartial—or who, like Croce, identify history with philosophy. The great bulk of what is presented in the works of standard historians is fact independent of what may be interpolated suggestions, probabilities, or hypothetical linkages. And it seems an exaggeration to assert that this hypothetical element escapes all crucial testing, even though consonance of its consequences with present fact is not enough to verify a given hypothesis as the only possible one. We have been hearing much of late from historians as to the undesirability, and indeed the impossibility, of unbiassed or disinterested historiography. Perhaps the example was set by Mommsen's saying that "history is neither written nor made without love or hate". It is followed in Creighton's remark that history "without so-called prejudication is a mere illusion", and in similar utterances of Cambridge historians to the effect that a historian's impartiality is an unmeaning affectation such as would paralyse judgement, or that history must be conceived from some definite standpoint. But if such statements seem somewhat paradoxical to the layman he may discount them by a charitable interpretation, taking them to mean that historiography is necessarily to some extent an art though not necessarily an exercise of arbitrary prejudice, of deliberate or uncontrolled partisanship, of surrender to passion, or of *suppressio veri.* What is implied seems rather to be that the historiographer, if he is to rise above commonplace or dry-as-dust narration, must have some insight or inspiration to see, and must speak not merely as an accurate scholar but also as the spokesman of his age or his people. As Lord Acton expressed it, "a great man may be worth several immaculate historians". With that statement we can all agree. And if it be said that then a historian's judgements are but documents in the history of ideas for later inquirers to study, or that they throw light on the historian's mentality rather than upon the subject-matter of which he treats, it may be retorted that natural science is also mainly the outcome of what were once the hypotheses of individuals. Without being characterised by certainty at any stage or in any particular production, historiography, for all its hypothetical ingredients, may make progress towards the truth. Even the reconstructions of history in the light of one or another broad speculative theory, such as those of S. Augustine, Herder, Condorcet, Hegel, Comte, and Buckle, if all largely superseded, were not all wholly imaginative. In short it bespeaks some misconceiving of natural science, which with all its objectivity is a survival of the fittest hypotheses and evinces certain characteristics of an art, when it is maintained that it is precisely the subordination of history, as research, to historiography, as art, which prevents history from being regarded as a science.

Perhaps the main conclusion to which this discussion points is that comparisons which have been drawn between history and science, with a view to belittling the former of these departments of knowledge, have generally been vitiated by contrasting the one of them at one level of its organisation with the other at another level of its organisation and developement. In historiography at its highest level we catch history passing over into philosophy, as is instanced in the words of Prof. Bury: "What is the position of modern history in the domain of universal knowledge depends in the first instance on our view of the philosophical question at issue between idealism and naturalism". Yet historiography, at this its most speculative level, has been disparaged by contrasting it with science at the level at which it is but humdrum process in nearest touch with concrete facts. And history in its incipient stage, as an endeavour to establish bare facts, has been decried by instituting a contrast between it and science at its highest reach of abstractness, when it presents us with rigid and universal, if hypothetical, law. Such treatment of history is unfair and unilluminating. If, on the other hand, we recognise how dependent historical investigation, for all its concern with the particular and unique, is on science of the common, and how dependent science, such as abstracts from the particular, is on the historical—in the broader sense of the word—we find the hard lines which some have been anxious to draw between them already beginning to soften. If we compare and contrast these departments of knowledge, both as sciences and as arts, at their corresponding levels or stages we observe no thorough disparity between them. We certainly find marked differences, in more than one respect. But whether these differences justify the refusal of the name of science to history depends, in the last resort, on where we agree arbitrarily to draw a line through the continuous body of knowledge which, commencing in natural history, issues in applied mechanics: a line on the one side of which all is science and on the other side of which all is to be called by another name.

Having concluded a brief survey of the relations in which human history stands to the sciences, and to knowledge in general as psychology and logic compel us to describe it, I will proceed to deal similarly with another department of presumptive knowledge or belief which is dependent on human history and closely connected with the historical: viz. dogmatic theology, or the field of revealed religion.

In thus separating dogmatic theology and revealed religion from philosophical theology and natural religion I respect the distinction between the two which has generally been recognised since it was explicitly drawn in the age of Aquinas. Several theologians have recently maintained that this distinction is arbitrary, superficial, and misleading; but they would seem only to have pointed out that the line of demarcation is not at all points hard and fast. That is certainly the case. Nevertheless there are broad differences between the two fields which cannot be explained away. For instance, though natural theology involves a revelation, natural religion is a response to that divine utterance which the theist reads in the constitution and order of the cosmos, the developement of human nature, its rationality and morality; while revealed religion is mediated by unique events occurring at particular dates and places, and by the life-histories of individual persons to whom divine inspiration is ascribed. Natural theology, again, is naturally discovered truth, or normally attained belief, based on science and philosophy; while revealed theology has generally been regarded as truth communicated 'ready-made', so to say. Natural theology is consequently comprehensible by normal human faculties; but some of the doctrines of revealed theology assert as truth what is not only not naturally discoverable but also not wholly comprehensible by the natural understanding, working with conceptions involved in, or derived from, ordinary knowledge. Further, the faith which is exercised in natural religion is but a further venture beyond proved knowledge, as logic would describe it, which is in essence continuous with the faith involved in all scientific induction: it is the active trust in the reality of the unseen and hoped for, but objectively suggested, while as yet unverified or purely ideal, which in the field of science often issues in discovery and practical invention; whereas the faith involved in the accceptance of revealed dogma, as in the *credo* of Anselm's *credo ut intelligam*, is the relatively passive assent to doctrine on the ground that it is proffered by external authority. Such differences as these authorise the separation of dogmatic theology from natural theology, and indicate the close connexion of the former of these departments of thought with human history.

There are of course several revealed religions, in the technical sense of the phrase, but it will suffice to refer only to the one in whose atmosphere we live, and to consider the relations in which Christian dogmatics stands to knowledge of other kinds. And whether the system of ecclesiastical doctrine which commands the assent of all the churches, to speak of that alone, can be called a science, or even a knowledge, is partly the same question as whether historiography is a science or a knowledge. For, in the first place, dogmatic theology uses for the systematisation of its data the same guiding and connecting idea as modern history, viz. that of developement. And, in the second place, the body of patristic doctrine rests on the historiography comprised in the New Testament, which was confidently and, in the modern sense of the word, uncritically taken by the fathers to be unadulterated history and divinely communicated truth. Thus the truth-claim of dogmatic theology stands or falls not only with the trustworthiness of the historiographical method in general but also with the trustworthiness of the particular products of that method which are enshrined in canonical documents.

The first matter to be considered is the nature of the data on which dogmatic theology is based. As we are not eye-witnesses of the events nor hearers of the utterances on which Christian doctrine rests, the data that are given to this generation and received by it are different from those afforded to the persons amongst whom the founder of Christianity lived. Between them and us intervene oral tradition and writings. The mental processes involved in the construction of the oral tradition are for ever removed from our direct observation, so the written records contain the whole of the data that now admit of examination. Further, the historical occurrences which evoked religious experiences and were the foundations of doctrine became data for theology, over and above history, only when disciples answered for themselves the question "What think ye of Christ?" or when His acts and utterances were overlaid with an interpretation of His personality. Thus, although Christianity, as a religion and a system of doctrine, is undoubtedly founded on history or the historical, what for us now are its foundations and data contain also the historiographical. If one may be allowed the bull, historiography preceded written records, while for our generation the forthcoming data are historiographical in the literal sense. Historiography, as we have seen before, and as its representatives in this University have been foremost in proclaiming, generally involves somewhat of art, or of preconception and of emotionally or otherwise motivated constructive imagination; and when concerned with the acts and utterances of a pre-eminently impressive personality, around whom the most soul-stirring hopes and sentiments have centred from His day to our own, it is liable in an exceptional degree to be pervaded with interpretative beliefs. And this fact has but comparatively lately come to be reckoned with by Christian historians and historical critics. The student of the New Testament at the present day, though equipped with the apparatus of textual, historical, and higher criticism, cannot always, by those instruments alone, arrive with certainty at history free from its historiographical factor. He may conceivably be able to ascertain which was the original text and the meaning (for its author) of its language, to eliminate later interpolations, and even to prove that a given interpretation of the facts as they simply happened is the most primitive that we can trace. But when all this has been done it remains a further question whether the primitive interpretation is the true, or the most probable, interpretation. The impression produced by the personality which they portray was doubtless the cause of the gospels, not they the source of the impression; yet the impression implies subjective wax as well as objective seal, and in turn has its presuppositions. Thus the problem involves a question which lies outside the sphere of historical investigation and criticism. History may decide how many interpretations the scriptural records contain, which of them is the most ancient, and how it underwent developement; but the further question is whether the original interpretation is sound and its later explication issues in clear and self-consistent doctrine. From the days of the early fathers almost to our own time appeal has been made to what was apostolical: that was adopted as the criterion of truth. Modernity, however, cannot accept that criterion as final. Instead of antiquity, authority, œcumenicalness, and even practical efficiency for the commending of Christianity and the promoting of the religious life, the criteria of verifiability and consistency have been exalted to the highest place. And only in so far as these tests can be satisfied will dogmatic theology be henceforth allowed a place within the sciences.

The relation of revealed theology to history, and to the historical in the broader sense in which all sciences of the actual are systematisations of the historical, is of importance for a philosophy of the departments of knowledge. And it should be made plain that academic distrust of the historiographical factor in the historical data of this theology is not necessarily aversion from either historical facts or the historical method.

It is true that scepticism as to ecclesiastical dogma is nowadays based on considerations connected with history rather than with natural science, as was the case some decades ago. And this seems to be due as much to movements in theological thought as to change in the scientific habit of mind. There is indeed a modernist theology which would surrender the alleged history on which traditional Christianity professes to be based, and yet retain the ideas and developed doctrines of that religion. But the distrust to which I am now referring is not a thoroughgoing scepticism as to the methods of historical research, such as deems all testimony untrustworthy and all historical conclusions uncertain. It does not consist in the relinquishment of all hope that the arbitrary and conjectural elements that have often vitiated historical criticism can be winnowed away.

Nor is it the suspicion, cast out of science but subconsciously cherished by some historical critics, that Nature is so rigidly uniform that any breach of uniformity asserted in the gospel-record is necessarily fictitious. Nor, again, is it due to the tendency to decry local and dated events as too parochial and insignificant to be accepted as mediations of eternal truth, such as every man ought to believe for his soul's health. All these prejudices may be discarded, and one may even hold that the application of the historical method to the study of the New Testament has brought our generation more directly face to face with the founder of Christianity than any age since His own: and yet it has been found possible to question the necessity of that interpretation of His person which, appearing in germinal form in the earliest extant records, was developed into doctrines such as those of the Incarnation, the Trinity, and the Atonement. To the layman, unversed in the intricate subject of New-Testament criticism, the written sources, even when their respective contents are sifted and rearranged, do not appear so harmonious as to shew clearly that the interpretation, of which catholic doctrine is the explication, was the only one forthcoming in the apostolic age, or that it is compatible with *all* the historiographical data, or that it is an essential presupposition of the joy and peace which the first Christians evinced. It is not necessary to enter into discussion of these questions; but the fact that such questions can arise is one among the considerations which led me to submit the academic assertion that the historiographical data of dogmatic theology are not of such a nature as to allow of that body of beliefs being regarded as certainly a science, or as a department of knowledge, save as knowledge concerning the history of thought. By a science I mean a systematisation of knowledge, or probable belief, based upon indubitable or verifiable fact; and as the original interpreters of their own experiences are, in the present case, not accessible for cross-examination, the data of dogmatic theology cannot be verified as can those of a science, strictly so called. The most we can now do with these data, with a view to providing a foundation for knowledge, is to select only those into which questionable interpretation does not enter.

This line has of late been adopted by what is known as the liberal school of theologians. Discarding the dubitable, this school would see the essence of Christianity in its founder's teaching, i.e. His ethical theism, and its manifestation in His life. Their view has been criticised by the conservative school as one that is not the outcome of historical inquiry but of theological prepossession—the prepossession that Christ resembled a modern protestant moralist. And certainly the view is not the outcome of historical inquiry in the sense of being dictated by the historian's original records, even when these are sifted by criticism using objective standards—which is what seems to be meant. But it is the outcome of historical inquiry in the more fundamental sense of discrimination between history and historiography, which conservatism is content to identify.

The liberal theologian, however, has gone further than discarding some of the alleged data on the ground that they are intrinsically dubitable in that their origin is in historiography rather than in history. He has inquired whether the explicated doctrines in which the original interpretation has issued are consistent with one another, with the totality of the historical, or even the historiographical, data as well as with a selected part of them, and with the more assured and verifiable knowledge with which these doctrines can be brought into relation. And these inquiries have encouraged his belief that a philosophy of the sciences requires him to persist in a procedure such as those who accept the historiographical as historical have pronounced arbitrary.

The issue now shifts from data and primitive interpretation to the developement of doctrine, and into that field we must follow it. The phrase 'developement of doctrine' is of course metonymical. Doctrines are not like organisms, of which developement from an embryonic stage can be predicated with literalness: they do not live and grow in independence of human thought. The concrete fact of which the phrase 'developement of doctrine' is an abstract expression is the advance in the thought of one or more doctors upon the thought of others. Accordingly, the analogy with what psychologists call 'explication of the implicit' is, so far as it goes, more apt for the description of doctrinal developement than is the more common one offered by biology. The human mind only gradually comes to make distinctions, in many cases, where all the time there have been differences; the boy's implicit notion of his bowling hoop, for instance, only becomes explicit when he learns clearly to conceive of circularity, rigidity, and velocity. And developement of doctrine is, in the opinion of many theologians, but the gradually attained explication of what was more or less vaguely stated in the New Testament. The implicit, in the psychological sense, is not to be confounded with the logically 'implying'. Logical implication is a relation between different propositions, not between degrees of discernment or of discriminatedness. Deduction of the logically implied doubtless is involved in the complex process of doctrinal developement, but in the patristic age it was but subservient and instrumental rather than constitutive of developement. In the scholastic age it played a larger part. The method of Aquinas, e.g., critical as he could sometimes be of authorities, consisted in asserting, as premisses calling for no proof, sentences of Scripture, of a father or compiler, or of Aristotle, and, by application of scholastic conceptions and terms, deducing implications from them. Thus the gospel was expanded into the *summa theologica*, and the method was doubtless regarded as an explication of the implicit. But in the more formative period the doctors of the Church were rather employed in eliciting the meanings of original beliefs by appeal to continuous tradition; and general councils were not debating societies relying on human understanding as infallible. The developement which they promoted also included the removal of ambiguity, which is distinguishable from giving definiteness to the vague. Thus the resort, at the council of Nicea, to non-scriptural terms in formulating a creed was justified by declaring that it had become necessary to fix and safeguard the one true meaning of ambiguous scriptural phrases, and to find expressions for it such as Arianism could not wrest or interpret heretically.

Developement of doctrine has already been analysed into three processes; and they all presuppose not only such facts of history as are not reasonably deniable but also the truth of a particular and traditional interpretation of some of those facts. Should the historiographical data involve any legendary matter, any literalisation of metaphor, or any ascription to the founder of Christianity of what He did not say or mean, so far will doctrine developed from those data be Christian only in the sense that it is accepted by Christendom, and not in the sense that it is doctrine for which Christ stood; and—what more nearly concerns us here—it will not necessarily be science or truth. And this doctrine will be further vitiated if, in the course of explication of what is implicit in the received historiography, extraneous ideas and assertions, especially if incompatible with the more certain data, have been introduced from non-Christian sources, whether by apostles or by later theologians, and incorporated in ecclesiastical dogmas and institutions. Developement will then include a fourth process. It will consist partly in assimilation or absorption from without, and no longer in pure explication of what was within. And at this point it becomes necessary to make use of biological analogies in order to describe doctrinal developement completely.

It will be useful, by way of fixing an ideal limit and a standard of reference, to mention a conceivable view as to the system of ecclesiastical doctrine which has never been actually held, viz. that it is the outcome of no developement. There was some approximation to this view in the first centuries. The conceptions of orthodoxy and authority then prevalent found relatively little room for the notion of developement. A body of doctrine was supposed to have been divinely delivered to the saints, to be transmitted as an heirloom without mutilation or addition. The function of a general council was conceived as not to discuss questions as if they were open, save in so far as to shew up the errors in heretical opinions, but to state in what way questions had been for ever closed. Novelty was the mark of heresy, and apostolicity was the criterion of orthodoxy. Such views are embodied in the Vincentian canon, *quod semper*, *quod ubique*, *quod ab omnibus*. Rigorously taken, this canon would exclude developement and imply the static completeness of doctrine from the apostolic age. And one forthcoming conception of developement does not essentially differ from this implication. This conception is derived from the obsolete biological supposition that the parts of an organism exist as 'preformed', though on a minute scale, in the embryo, and that the developement of an embryo into an adult consists simply in enlargement of these parts, or in a redistribution of them in space, as in the case of pulling out the tubes of a collapsible telescope. Conceived in terms of this analogy, doctrinal developement should consist in verbose paraphrasing of concise statements, without refinement of definition. But such substitution of prolixity for terseness is not even explication of the implicit, and mere unfolding of the preformed is a description of non-developement rather than of the developement which has actually taken place. If biology is to supply a serviceable analogy, developement in doctrine must be conceived in terms of the modern notion of evolution, or as epigenesis. According to this conception, the organism is not an educt from the germ, but a product of interaction between the germ and its environment. Growth into the adult form is growth *out of* the embryo, not mere enlargement of it. The growth involves assimilation of matter from without, so that what the germ grows into was never wholly present in it. An acorn, in isolation, is not a potential oak: it has not the potency to become an oak apart from absorption of external matter. If the developement of doctrine involves growth and assimilation similar to that in virtue of which the seed becomes a tree, a fact is indicated which is disturbing to conservative theology. Less of unchangeableness than orthodoxy would welcome is suggested. Hence Newman, in his theory of the developement of doctrine, endeavoured on the one hand to make use of the biological analogy and on the other hand to maintain that the dogmas of modern catholicism were possessed, substantially as they now are, by the Church from the first. And such growth and transformation as he admitted is not free or natural, but owes its alleged immunity from error or alienation to the control of infallible authority—which, of course, is itself a product of developement.

Dogmatic theology, then, can only give an account of itself by adopting the guiding notion of epigenetic evolution; and recourse to that notion exposes the system of ecclesiastical dogma to the charge which the liberal school have urged against it. This school asserts that developement has consisted partly in the appropriation of inapt, unclear, and now obsolete metaphysical concepts, and partly in the absorption of alien and questionable matter. Such infiltrations and secularisation, it is said, ministered to the maintenance of the Church and to its influence as an institution, but at the same time to progressive obscuration of the primitive simplicity of the gospel and of the ethical theism which the founder of Christianity taught and lived. It is not now necessary to inquire into the grounds on which this charge is preferred; but it may be concluded, from our examination of the developement which dogmatic theology presupposes, that while the process of developement *may* have introduced more of the doubtfully true into the system of doctrine, it can hardly have effected the elimination of any of the doubtfulness which attaches to the primitive data in virtue of their historiographical nature. In so far as dogmatic theology is held to rest on other foundations than those with which history is concerned, its claim to be a science, or a body of knowledge based on facts, needs consideration in another context, viz. in connexion with religious experience as it is enjoyed by Christians of the present age. That is a question belonging to psychology and theory of knowledge rather than to history, and it shall receive discussion when theology in general is dealt with in a later lecture.

Lecture V

THE RELATIONS OF THE NATURAL AND THE PURE SCIENCES TO EACH OTHER, AND TO PHILOSOPHY AND METAPHYSICS

THE FIRST question which I propose to discuss in this lecture is the relation in which the natural sciences, regarded collectively rather than individually, stand to metaphysics. And this question can best be approached by observing the significance of the fact that science issues from common-sense knowledge, or presumptive knowledge. It is at the level of the common-sense organisation of individual experience that our knowledge of the things or the phenomena which the sciences connect and explain, and of the facts which science takes over as its initial data, is constituted. And at this level something more than the primary certainties, on which I touched in an earlier lecture, is involved. Between private percepts, of the here and now, and physical, public or common, and persisting things, of which common sense and science treat, there is a gulf which, if unsuspected by common sense, should be obvious to the psychologist. Common sense, in overlooking the difference between the apprehension of the one and the other of these kinds of object, tacitly assumes a theory of knowledge. And this theory of knowledge is the one with which, in the first instance, science set to work. It involves the supposition that a public and permanent thing, such as the sun, is apprehended with the same immediacy and certainty as is the private percept of a flat patch of yellow. Indeed such common objects are generally called—I would say miscalled—perceptual. They are also deemed to be real, not only as distinct from imaginary objects, but also in the sense that apprehension of them involves no supposition, or no contribution from 'the mind itself'. Some philosophers uphold this view. Science, however, at its more advanced stages, repudiates this realistic theory of knowledge of the external world. And the psychology of cognition reveals its discrepancy with certain empirical facts. Indeed it seems plain to me that of all the forthcoming theories of knowledge, such as realism, idealism, positivism, and phenomenalism, the last is forced upon us, and the rest are refuted, by facts of which any theory of knowledge must take account. We are now concerned only with the realism which science takes over from common sense, and the refinement to which science subjects it in passing on to what may be called scientific realism. The former and cruder theory asserts that our bodies, and all the physical processes which physiology tells us are involved in perception, simply make the perceiving, the *perceptio*, of a thing possible, but do not in the least determine the nature of what is perceived—the *perceptum*. Perception, that is to say, is affirmed to be comparable to vision through plate-glass; it involves no chromatism or phenomenalising, so that no distinction is to be drawn between reality and appearance. According to this theory all the different shapes which a dinner-plate presents to percipients looking at it from different positions, and all the colours which it has when seen in different lights, should be alike real. Equally real should be all the queer things seen when one presses one's eyeball, when one dreams, or is delirious, or is under the influence of drugs. The indefinitely vast complexity which, if this theory were true, would characterise our world, should render that world many times less amenable to physical science than it is. Accordingly we are not surprised to find that even common sense does not retain a realism of this type, but rather turns to another, viz. to what has been called the causal theory of perception. It holds, that is to say, that the real object, such as the sun, causes not only an individual's *perceptio* of it, but also his particular percept or *idion*, which may differ in quality from the percept of another individual looking at the sun. The percept is now taken to be an appearance of the real sun; and realism has partly lapsed into phenomenalism. Science encouraged this lapse when, adopting the distinction between primary and secondary qualities, it regarded extension as real but colour as subjectively contributed. The physical object, as well as the *idia* of individuals 'perceiving' it, thus came to be regarded as a phenomenon of real, or ontal, thing; and this reality was supposed to be what physics describes in terms of insensible atoms, ether, electrons, etc. But even so much of realism as the assertion that our percepts are always caused by the real core in partly unreal physical objects becomes untenable when we take account of hallucinations, of 'sensations due to inadequate stimulus', as they are called, such as the light we see when we bump our heads in a dark room, and those known to psychologists as 'subjective sensations'. For in one of these cases the real object that should cause the perception, and thereby be perceived, is not perceived; and in the other two cases it is not even involved. Thus the causal theory breaks down. And once it is allowed that secondary qualities are appearances, or are effects partly of the primary qualities of external things and partly of our bodies, on our sensibility, there is no necessary stopping-place before complete phenomenalism is reached. It would be all one to science if the primary qualities of matter were as much appearances of something else as the secondary qualities have been allowed to be.

This complete phenomenalism fits all the facts of normal and abnormal sensation, whereas the realistic theory of our knowledge of the external world seems to be refuted by both physical science and the psychology of cognition. The public object of common experience is not an aggregate or a symposium of the immediate objects or *idia* of different percipients, nor are they parts of it. They are often not supplementary: the shapes of a thing which are actually perceived at different positions with immediacy and certainty are not compatible with what we call the real shape. Nor can it be pleaded that though, in the order of knowing, the physical object, or our knowledge of it, is constructed out of *idia* yet, in the order of being, it is prior and is the cause of our *idia* and of our perceiving them as appearances of it. For, as I have said before, critical inquiry renders the latter conclusion unwarrantable. There would seem to be no transition from an individual's private and transient sensations to the permanent objects of common experience save by intersubjective intercourse, or comparing of notes, and by addition of suppositions to our sensory *posita* and the relations that can be immediately and certainly observed to subsist between them. In short, the so-called perceptual objects, from which science sets out, are largely conceptual. They are what we have reason to *think* exists, in order to correlate and account for what is thrust on our senses, and to make, out of the chaos of sporadic diversities, a rational world—i.e. a world of entities persisting as identities in time. These objects, then, are not to be taken for items of ultimate reality. Yet, I would insist, they must be some version or function of the real, else the science which uses them as indispensable standardised constants would hardly be consistent and valid, or even forthcoming.

The two main conclusions to which I have thus far been led are (1) that, in the facts from which science sets out as if they were of bed-rock nature, there is already an element of suppositional and rationalising venture, which is justified only by its pragmatic success, and (2) that the forthcomingness of physical science involves that distinction between the ontal and the phenomenal —not between two worlds, but between the real or ontal world and its appearance to our minds—which other theories of knowledge would, in different ways, dispense with. The error inherent in the realism which science initially and uncritically takes over from common sense in no way debars science from being a (phenomenal) version of truth about the ontal, and does not vitiate science, as such; but it becomes a matter of importance when the precise relation of scientific to metaphysical truth is in question. Science, studied in aloofness from what may be called its higher criticism, may foster confidence that critical problems do not exist for it, and so engender superstition or obscurantism.

Diverging for a moment from the metaphysical presuppositions that are intrinsic to science in virtue of its continuity with common sense, I may observe that the truth that our primary scientific facts are partly theory throws light on two eliminations which science makes, thereby constituting itself a kind of knowledge distinguishable from other kinds with which we have been already occupied. Science abstracts from the subjective factors involved in knowing, and regards its objects, though it calls them phenomena, as if they were as independent of all subjects as they are of any one subject. This is a fruitful device, indispensable to science; but, philosophically regarded, it is a fiction. Thus physical science is sharply marked off from psychology, the subject-matter of which science presupposes but ignores, and from philosophy, which cannot wink at this elimination of subjectivity, or treat science as if the elimination had not been made. Again, science abstracts from certain aspects of its objective data. It finds identities for thought by ignoring, as negligible *ad hoc*, what are diversities for sense. In passing from facts to generalisations and laws, science necessarily confines itself to the repeatable and common, which are in some degree abstractions. Hereby science is differentiated from history, though again the historical is presupposed. And in so far as science isolates itself from history in order to pursue its own special business, it may be said to present us with but a diagram or a skeleton of Nature. In other words, science is precluded from supplying a living picture of Nature as manifested in the presentational *continua* of human beings as well as from describing Nature as a sum of interactions between the known constituents of the ultimately real world. Science makes its repeatables out of the historical and unrepeatable by abstraction and by substituting concepts for percepts. The element of art thus introduced into science is as fruitful as it is indispensable; but unfortunately it is also the first step towards a vicious abstractionism which has sometimes been confounded by natural philosophers with the metaphysical method.

As I was observing just now, science sets out with certain metaphysical presuppositions which it took over from common sense. They are revealed by epistemological scrutiny, which is one of the chief functions of philosophy. The plain man who does not exercise this scrutiny is unaware of the tacit assumptions which his common-sense knowledge involves. He is at home in such 'knowledge' as he has, and all seems plain in it because his eyes have not been opened to the difficulties with which critical philosophy finds it to be beset. Consequently philosophy seems to him to raise obscurities and to create vain fancies where knowledge had seemed to be clear and sure. Similarly metaphysics is wont to be most vehemently scorned by those whose mentality is most deeply steeped in unsuspected metaphysics, and who are unconscious that they are talking metaphysics of their own while railing at the metaphysics of professionals. One needs to be a philosopher of some standing in order *not* to be a metaphysician. And this explains why scientific thought, issuing in certain doctrines of matter and energy, in conservation-principles, and in a mechanistic world-view, was profoundly metaphysical until it had pursued to some extent that self-examination which it began to undertake about fifty years ago, and the results of which are being expressed by its present representatives.

It will perhaps repay us to recall the changes which have recently taken place in the attitude of science towards metaphysics and philosophy. In the latter half of the nineteenth century some of the chief spokesmen of science sought to commend its unique excellence by disparaging the metaphysics of the more pretentious schools of philosophy. And in this they were not without provocation. For instance, one eminent post-Kantian idealist had declared that "the philosopher performs his task without regard to any experience whatsoever, and absolutely *a priori*"; and another had derided "the blind and thoughtless mode of investigating

Nature which has become generally established since the corruption of philosophy by Bacon and of physics by Boyle and Newton". But reaction against extravagances such as these led some physicists, e.g. Tait and Maxwell, to consider themselves to have sacrificed dignity by so much as crossing the borders of metaphysics for war-like purposes, much as the ancient Hebrew regarded himself as rendered unclean by entering a country presided over by other gods than his. In an address contained in his *Scientific Papers* Maxwell alludes to the "den of the metaphysician, strewed with the remains of former explorers, and abhorred by every man of science". "It would indeed be a foolhardy adventure", he tells his audience, to lead them "into those speculations which require, as we know, thousands of years even to shape themselves intelligibly." Nevertheless, Maxwell went on to say, the cultivators of mathematics and physics (such as he was addressing) are led up in their daily work to questions the same in kind with those of metaphysics, but they approach them as trained by a long-continued adjustment of their modes of thought to the facts of external Nature. Maxwell evidently was convinced that the facts of external Nature, as taken over by mathematical physics, are devoid of any metaphysical factors, and that the physico-mathematical modes of thought employed in the scientific explanation of them are adequate to the solution of the quasi-metaphysical problems to which these facts lead up. And these implicit views were explicitly maintained by some of his contemporaries. Not only was it then commonly believed that science was positive knowledge, in the strictest sense, it was also asserted that the scientific method is the sole means of approach to the whole realm of possible knowledge; that there are no reasonably propounded questions that science could not hope to answer, and no problems worth discussing to which its method was inapplicable. Many representatives of science thus sought to identify it with the whole of knowledge and with all that could call itself philosophy. Metaphysics, other than that which consisted in taking physics to be ontology, was deemed impossible, and natural theology was accounted superfluous. From physicists and biologists such views spread to many psychologists and philosophers. Indeed for a time philosophy became as much enslaved to physical science as in the middle ages it had been enslaved to dogmatic theology.

Science has receded from the pretensions into which it rushed, as a giant rejoicing to run his course, when temporarily elated with its own success and its acquired prestige. And it may be observed that the plausibility of those pretensions was due to the prevailing supposition that physical science was independent of, and unrelated to, such departments of knowledge as psychology, theory of knowledge, and history. On the contrary, a philosophy of the sciences must insist that these latter departments of knowledge reveal the presuppositions of physical science, and set the bounds to its scope and functions. Only when its relations to them are appreciated can the true nature of scientific knowledge be apprehended.

To continue the more recent history of the attitude of science, that is to say of its representatives, towards metaphysics, it may next be remarked that contemporaneously with the hostility of Prof. Tait and others to metaphysics, and their endeavour to "sift the truth from the metaphysics" contained in traditional statements concerning physical conceptions such as those of matter and energy, science was claiming, on the strength of the realistic trend in its theory of knowledge, to offer a metaphysical substitute for the metaphysics of pure philosophers. The realities, of which sensible or molar bodies are appearances, were believed to be knowable to science, and indeed to be identical with the *mikra* in terms of which theoretical physics explains macroscopic, or directly observable, phenomena. The physics of this microscopic order has been developed largely with a view to finding a rigidly mechanical explanation of phenomena which *prima facie* are not mechanical. And it was taken in some quarters to disclose the real which underlies the phenomenal. Thus theoretical physics came to pass for ontology, or metaphysics of the physical world. The reality of the physicist's *mikra*, however, is not a condition of the forthcomingness of molar and empirical facts. It is not even a condition of their explicability, but only of their mechanical explicability. And intelligibility of that particular sort is no precondition of science, as was once commonly believed; it is but a human *desideratum* and a luxury, the craving for which perhaps bespeaks an aesthetic element in the constitution of human reason. We can accord to the fruitful theories concerning hypothetical *mikra* the appreciation which is their due without taking figurativeness for literal truth. That Nature behaves *as if* the ether, the electron, the nucleus and revolutions round it and all such microscopic machinery were realities, is largely fact; and that is all that we are entitled by experimental science to assert, and all that such science need demand. That these microscopic entities, which are neither observed nor inferred, mediate metaphysical knowledge, pure and undefiled, is, in the light of the past history and of the present internal inconsistencies of theoretical physics, even more precarious than it is superfluous. Science neither reveals nor presupposes a rigid mechanism of real microscopic entities. Schemes involving thoroughgoing continuity and schemes presupposing radical discreteness are, at the present stage of scientific speculation, as serviceable as they are incompatible. The ultimate elements into which the theoretical physics of to-day seems to resolve the world are neither ontal nor phenomenal: they constitute a fictional or symbolical descriptive scheme, or rather an incongruous set of schemes, partially applicable, but into all or any of which even inorganic Nature refuses wholly to fit. But it is not necessary at this hour to elaborate indictments against scientific realism, nor even to reproduce the grounds and reasonings on which epistemologists have based them. For the view that microscopic physics yields a metaphysic of Nature is refuted and abandoned by the present generation of physicists. The scepticism with regard to it, which until lately found expression only in a *savant* here and there, has been affirmed by several eminent authorities to have now become a characteristic of scientific mentality. Indeed Sir James Jeans has recently declared that the outstanding achievement of the physics of the twentieth century is not the theory of relativity, or the theory of *quanta*, or the dissection of the atom, but the disclosure that science has not reached ultimate reality. But in the form in which this disclosure had previously been made by certain philosophers it includes the wider assertion that the abstractive method of science never will, because it never can, reach ultimate reality: the very nature of the scientific method of explaining diversity by reducing it to identity, in virtue of which science tends to issue in mathematics, renders this feat intrinsically impossible. In straining out the sensible, the qualitative, or the historical, theoretical science strains out the actual and the real also. It may arrive at valid laws and equations, but not at the realities for which its symbols stand. It may reach the conceptual, or a system of abstract ideas; but an abstract idea is not an ultimate reality: metaphysically it is nothing. If so, the replacement of the mechanical model, which the science of the nineteenth century favoured, by the mathematical symbol will not enable science to pass for an improved metaphysic, nor for metaphysics at all. When Sir James Jeans goes on to say that Nature is written in mathematical language, implying that Nature is pure thought, he would seem to be propounding a doctrine which is further from the truth than scientific realism is.

We may conclude, then, that science of the microscopic order is not genuine metaphysics, or a substitute for ontology. By 'metaphysics' I mean ontology, together with the department of epistemology that is concerned with knowledge of the ontal realm. I do not mean *a priori* treatment of cosmological or other problems, though some philosophers would define metaphysics so; for I do not believe that the *a priori* method alone can yield truth as to the actual world. Metaphysics that is not based on science can be but a pastime if the ontal reveals itself only through the phenomenal or the historical, and if the element of 'brutality' in the data or the analytica of all knowledge be determinative. But though metaphysics presupposes science, in the order of knowing, science is not ontology. If science claims to be knowledge of the real world it uses the word 'real' with the meaning which it bears in ordinary parlance, but not with that which the word bears in ontology. Science presupposes an ontal order—she has never been idealistic—and an order which has a structure enabling reason to find rationality in the phenomenal; but, as to the nature of the ontal, science is agnostic.

I may now return to the body of such scientific knowledge as has never claimed to be metaphysics, i.e. to pass through the phenomenal veil into the sanctuary of the ultimately real. The sciences of directly observable facts were said before to be characterised by a kind of systematisation which distinguishes them from other departments of knowledge, such as history on the one hand and pure sciences on the other hand. History gives system to its once-occurring facts by tracing chronological, causal, and genetic connexions which subsist between them in all their concreteness and uniqueness; whereas the natural sciences consider such facts only as instances of general laws and as members of classes, or in respect of their common and repeatable qualities. Consequently science begins in comparison and classification, and passes on, by way of generalisation and induction, to establish laws as to sequence and concomitance. So much of procedure is common to all the different sciences of Nature and, roughly speaking, constitutes the scientific method. Perhaps the broadest definition of the distinctive characteristic of science is that science is an endeavour to render the world intelligible in a special sense. The intelligibility which science assumes the world to possess, and seeks to reveal, is not that of teleological reasonableness, but rather a quasi-logical rationality. More precisely, each science may be said to arrange its facts in a particular kind of apperceptive system, consisting of concepts drawn from its respective context of experience, or consisting of secondary constructions, the uniqueness of which lies in the condition that they must be capable of extending the sphere of knowledge, or of leading from the known to the unknown. To continue for a moment to speak of the different sciences before reverting to science in its generality, one may say that each specific natural science is chiefly characterised by the aspect of actuality which it is concerned to study, and consequently by its choice of what, in the diversity of the actual, it will account negligible in order to establish identifications, or partial identifications. Thus mathematics accounts the whole of actuality, and the totality of its qualities, as negligible for its own peculiar purpose; it makes a complete abstraction of quantity, and accordingly establishes the most perfect identification—the equation—that is compatible with mathematics stopping short of being one vast tautology. The most rigorously mathematical type of classical physics ignores every aspect of actuality save the movability of something or other that, unlike a number, has substantial or real existence as contrasted with ideal existence or existence-in-mind. Chemistry, as such, needs to recognise, for the prosecution of its own proper work, the qualitative diversity of its elements, even if it regards the carbon in carbonic acid gas as consisting of the identical atoms which constituted the carbon before its oxidation. What physics can dismiss as negligible, while establishing its identifications, chemistry must retain as indispensable. And so on, through the biological sciences to history, where quality and value count for practically all and quantity counts for practically nothing. But besides these different degrees of intelligibility of one sort, achieved by the several sciences, different kinds of intelligibility are also pursued, at least as makeshifts. If different data and different aspects of actuality may require different methods to render them intelligible in the same sense, they may also be made intelligible in other senses than in that of reducibility to identity or equality. And, as a matter of fact, this is chiefly what is attained in the more concrete sciences. Diversities of operations are called for in different departments of science, though one spirit or ideal animates them. In short, different sciences work with different explanation-principles and with different numbers of as yet indispensable categories. And each science has a right as well as a need to use such categories as its peculiar data may require for concatenating its facts and affording such explanation as is possible without missing the distinctive significance of those facts. Thus, while logic and mathematics require only the 'formal' categories, such as likeness and difference, numerical oneness and plurality or otherness, the science of motion involves the notion of substance, or of a continuant identity, such as cannot be immediately read off, like difference, from percepts. The other physical sciences need another 'real' or 'dynamic' category, viz. cause, if indeed cause and substance are not mutually implicative, or one and the same category, in so far as they represent anything actual. And these categories express relations which are not immediately discerned or read off, but are read into the immediately given. They are first postulated, then pragmatically verified. And they are postulated in order to make the data intelligible. The intelligibility to which they minister is not that which characterises the pure sciences: it is not transparency to the understanding, such as identity presents, but rather absorptiveness into an apperceptive system in which the familiar, however intellectually turbid, is the *explicans*, and where anthropic interpretativeness or analogising is substituted for direct apprehension. Here we encounter the first conspicuous manifestation of the activity, constructive rather than receptive, of human reason. The rationalising which is effected by means of the idea of causation, the irreducible core of which is the idea of an actual *conditio sine qua non*, is different from that which can be pursued in a pure science. The latter kind consists in identification: the former is, rather, interpretative assimilation. Still, the causal category is justified by its working; and, if suggested by reason, it is proved valid, and is consequently sanctioned or dictated, by the behaviour of actuality. Causal explanation is intermediate between explanation in terms of formal categories and explanation in terms of final causes (as in human history) or of such concepts as are required in biology, and is much more nearly akin to the latter than to the former kind of explanation. It is biology especially that reveals Nature as *naturans* as well as *naturata*; and in that sphere the attempt to explain the facts in terms of retrospective categories, or of a *vis* that is *a tergo*, alone, would be but to seek the living among the dead. We recall, in this connexion, the pitiable shifts to which Kant was put in his endeavour to stretch the categories sufficing for mechanical explanation of the inorganic world so as to cover the phenomena of life.

Thus the nearer any department of knowledge is to

embracing the richness and diversity of the actual world, the more numerous are the categories indispensable for it as a pursuit, or as engaged in research. And in the present state of our knowledge of Nature a science requiring more of such categories cannot be reduced to, or be subsumed under, another science requiring fewer categories, or pursuing but one kind of intelligibility or explanatoriness. If there should thus be irreducible diversity between some of the departments of natural knowledge, in spite of e.g. the partial applicability of physical methods in chemistry and of chemical methods in biology, science can never as a whole pass into science *par excellence*, or become the deductive system contemplated by Descartes. Such diversity would, for science *par excellence*, be irrationality, and would debar science from effecting a complete rationalisation of the world. The essence of reason—the reason which inspires science's chief aim and ideal—is identification as far as is possible. For complete identification, or for physical science to become as deductive as what are called the pure sciences are usually said to be, it is requisite that *causa* be identical with *ratio*, and that *causa aequat effectum* be either an analytical proposition or—if *aequat* means identity in all respects *save* spatial configuration—an assertion of the reversibility of physical changes.

But cause is not logical ground, and change is not reversible. Consequently it is impossible even to provide an *ex post facto* exposition of the deducibility of scientific knowledge such as happens to have been obtained by experiment and induction. Nevertheless throughout the history of science, regarded as applied logic, it is the deductive method, preferable on account of its certainty, that has been the ideal of science. Indeed knowledge, truth, reason, and rationality have generally been conceived in accordance with it. Thus scientific reason has been regarded as consisting in (1) immediate intuition of self-evident truths concerning actuality, or physical axioms, and (2) deduction from them of more particular laws and of facts. But, as for the axioms on which the physical sciences have been said to rest, it has been conclusively shewn, as it seems to me, that when they are not conventional definitions like those of the pure sciences, they are empirical generalisations, familiar enough to be mistaken for *a priori* truths. Thus the parallelogram-law, the principle of the lever, the principle of virtual velocities, and other such fundamental mediations between pure mathematics and physics as once were supposed to be independent of experience, have been found to be incapable of demonstration, and even of origination, apart from empirical observations. Newton's laws, on which Kant's *a priori* physics was based, are now recognised to be conventional definitions in a conceptual scheme or a pure science, suggested by empirical facts. Descartes professed to deduce his physics from an abstract formula, but his deduction consists in shewing that certain physical laws are compatible with that principle, not that they are unique consequences of it. And the abstract principles from which other rationalistic physicists derived their several systems of deductive physics admit of a plurality of particular consequences, while the one which each thinker took to be *the* necessary consequence was but that which the empirical facts required. However, it is hardly necessary to-day to elaborate arguments to prove that physical science is not *a priori* and deductive knowledge, or that its premisses are not eternal verities independent both of actuality as known and the knowers of it. And if the psychology of knowledge which I submitted in earlier lectures of this course be not altogether unsound, science is only possible through anthropic interpretation of historical data: all so-called description, in terms of law and logistic conceptions, ultimately presupposes interpretation.

Consideration of the alleged deductiveness of physical science has brought within view the sciences of logic and mathematics, and I will now proceed to state briefly what seems to me to be the relation in which these departments of knowledge stand to the natural sciences on the one hand and to philosophy, or, more particularly, to metaphysics on the other hand. I trust I am not deceived in thinking that one can venture to discuss such general questions without being either a mathematician or a symbolic logician.

Logic is wont to be described as the system of rules according to which we ought to think. Substituting for the word 'ought', which has an ethical insinuation, in this description, it would seem to mean that logic is the system of rules according to which we *must* think if, in so far as inference alone is involved in our thinking, we would avoid error and contradiction. As thus described, logic has no concern with the validity of premisses, nor therefore of conclusions, but only with the implications of propositions, be they true or false. It will stand in no such relations as science does to the historical, and will be but instrumental to science. It should be independent of psychology, which describes how we actually do think, rightly or wrongly, about things, and also of theory of knowledge. Thus it should take no note of the difference between analytic and synthetic propositions, or of the different interpretations put on '*S* is *P*' when that form of proposition is regarded from the points of view of intension and extension respectively. Hence it appears that the description in question is too narrow to cover the whole of the subject-matter discussed in text-books on logic, even of the deductive kind alone. And expert opinion seems not to be unanimous as to whether a logical system, or the 'theory of logic', can be consistently expounded without some 'epistemic'—i.e. psychological—elements being involved; indeed doubts have been expressed as to whether deductive logic, of the exclusively extensional kind, is an entirely pure science. The notion of implication, for instance, raises a dilemma. If implication, which logisticians declare to be indefinable, is reducible to inclusion merely, or to identity, a logical system should be a systematic tautology—as one eminent logistician affirms to be the case. If, on the other hand, logic enables us to advance to new truth, deductive inference should be an explication of the implicit involving some constructive activity of reason over and above the more passive intuition of identity. And then it becomes a question whether this activity involves mental operations presupposing empirical observations and memory thereof. Moreover the fact that in logical systems resort has been made to indefinables, and to descriptions of them by means of giving examples, has seemed to some critics to involve ultimately the appeal to experience which a pure science forswears. However, it behoves a layman to abstain from more than mentioning, for what they are worth, these suggestions that deductive logic does not, in its operations, keep itself so wholly "unspotted from the world" as it professes to do, but unconsciously imitates the behaviour of empirical things, only with more subtilty than professedly *a priori* physics has evinced.

Leaving, then, these disputable questions, and taking logic in a more comprehensive sense than that of a mere system of rules for deductive inference, I will briefly refer to the relation or the want of relation between the two main types of forthcoming logic, on the one hand, and science or scientific reasoning, on the other hand. The first of these types is the Aristotelian logic, in the expanded and refined form to which it attained in the modern period. It includes a doctrine of definition and of predicables, a metaphysic of substance and attribute, grammar, etc., as well as an element of exact science such as prepared the way for the developement of extensional logic and of algorithmic, or symbolic logic.

In this Aristotelian amalgamation of knowledges there are doubtless products of the confusion of the intensional point of view (from which *S* is *P* means that *P* is included among the attributes constituting *S*) with the extensional point of view (from which *S* is *P* means that *S* belongs to a class characterised by *P*). And it is the inexact or impure logic of intension in the Aristotelian scheme that has been abandoned during the progress of logic to an exact and quasi-mathematical science. Yet, as many eminent logicians of various schools have observed, it is this element in the old system which lends itself for the justification of the procedure of science, as research, and which ministers to that rational explanation which science pursues. And this is so because the intensional side of Aristotelian logic, like science itself, arose out of common sense, and common sense—as I have previously maintained— arose out of the human desire for rational explanation. Extensional logic leaves the attributes of a logical subject unconnected, as if their coexistence were a fact of chance, and it supplies no ground for preferring any one possible classification to another. It may serve to introduce an order into knowledge when the knowledge has been obtained, but in the acquisition of new knowledge it plays no part. Thus artificial classifications are generally useless to science, whereas natural classifications, ultimately based on the coherence of attributes, are fruitful. Scientific knowledge and explanation can only be obtained by acting on the assumption that the world is rational, or that there is a necessary coherence between the attributes of a thing; and that is assured by the common-sense metaphysics presupposed in the intensional interpretation of *S* is *P*. The essence of reason, certainly of the scientific reason, is the endeavour to see the contingent as necessary. I have previously made use of the assertion that scientific reason consists essentially in finding identities or making identifications. And the meeting-point of these apparently different or unconnected assertions would seem to be the analytical judgement—the only kind that in a purely logical sense is certified—in which *S is P* means that *P* is necessarily an attribute of *S*, and in that sense is included in *S*. The reduction of the synthetic proposition to an analytical proposition, which alone is metaphysically, scientifically, and logically certain, expresses the inmost desire of the rationalistic philosopher, such as Wolff; and we may recall Leibniz's faith in the necessity of what for us are contingent truths though, as he confessed, their necessity, or the reason of them (i.e., the inclusion, in their case, of *P* in *S*) can only be discerned by the mind of God.

The Aristotelian logic thus illustrates, and perhaps further elucidates, the assertion which I have already made on other grounds, that science, in so far as it is discovery of new truth, presupposes interpretativeness involving faith as to the metaphysical. Reason, in order to acquire knowledge of Nature, needs to regard some attributes as more essential than others: not, indeed, in the sense in which Aristotle's essences differ from his accidents—for science implies that all qualities are equally coherent and conditioned with the same necessity—but that some lend themselves better than others to the thought that penetrates into Nature's constitution. In other words, science needs to be selective, and extensional logic cannot recognise or justify its selectiveness. Again, in induction it is not always the number of instances, but often the 'weight' of few or of one, that compels us to a probable conclusion. And the validity of all conclusions reached by problematic induction is, as we know, dependent on indemonstrable assumptions, which are metaphysical because concerned with substances and causal conditioning.

The extensional logic which has been developed into logistics, or symbolic logic, loses in relevance to scientific reasoning and research in proportion to its gain in clearness and exactness. It is computational and expositional rather than normative and concerned with constructive thinking. This is, of course, no defect, because logistics renounces all connexion with theory of knowledge and is indifferent as to whether its laws, etc. apply to actuality. It also does not profess to shew how the science of mathematics, with which it regards logic as identical, has been built up, but only to give, after the event of discovery or construction, an exposition of the deducibility of mathematics from the basal and indemonstrable principles of the logistic system by means of the logical calculus. The influence of this newer logic on philosophy has been slight. But one observes that, in quarters in which the metaphysically imbued interpretativeness of human reason (the source of all our scientific knowledge) is disparaged, the logical constructions of the symbolic logician have been offered as adequate substitutes for interpretative categories such as that of substance. Such 'description' as then takes the place of explanation, however, usually ignores or misses what is significant for science through parsimony as to the metaphysical or the interpretative, which suggests the adage "penny wise and pound foolish". Also we have been told by some of the expounders of symbolic logic that this science, though pure, deals with entities which determine the actual world, and that therefore it cannot be ignored by a philosopher who would be other than an amateur. But though these entities are something and not nothing, they cannot be taken for substantial realities or for functional actualities, or be classed with the entities with which metaphysics and natural science respectively deal, save by a dogmatic fiat. The pure sciences are concerned with the realm of the valid, or the realm of subsistent relations which constitute the difference between truth and error in reasoning of the mathematical kind, and coerce our minds when they pursue such reasoning. But this realm is one of relations between ideal or non-actual entities, and there is no reason to believe that the ideal, which is abstracted from the actual, is the ontal. Nor is there any reason to believe that the entities with which the pure sciences deal determine the actual in the sense that it could not exist unless they were its form and order. On the other hand, we have reason to believe that the order and form described by the pure sciences, in so far as it is found by experience to be embodied in actuality, is brought about by the nature and behaviour of actuality, which conceivably might have been other than it is.

In symbolic logic and mathematics, then, we are not constrained by timeless 'realities' independent of our minds, but by the logical necessity of relations which subsist between idealities. The simpler numbers are doubtless got by abstraction from actual things, and other mathematical entities such as imaginary numbers, spaces of various numbers of dimensions, etc. are not discovered but volitionally created, while what are called axioms are postulated conventions, comparable to the rules of a game. We are only coerced by mathematical reasoning when we have adopted certain definitions and committed ourselves to certain axioms, while we were free to accept others.

But as to mathematical reasoning, such as has brought about the advance from simple arithmetic to the several departments of higher mathematics, its alleged deductiveness raises a dilemma similar to that which we encountered in logic. Rigour in deduction and progress to knowledge once undreamed of seem incompatible. And as mathematics cannot be regarded as a vast tautology, it is its deductiveness that has been called in question. Then, however, the difficulty presents itself, how it is that the theorems of mathematics, unlike the inductions of concrete sciences, appear to be demonstrable in the most rigorous sense. So constructiveness has been attributed, in the place of deductiveness, to mathematical reasoning: one recalls, e.g. Kant's synthetic judgement *a priori*, and Poincaré's 'mathematical induction'. I have before remarked that much of what Kant took to be *a priori* is to be suspected of being partly *a posteriori* and partly interpretative supposition, and I have alluded to mistakings of elements of that nature for the *a priori* in the sphere of physics. And now I would observe that there is much to be said for the view that mathematical reasoning, of the constructive kind, is essentially akin to scientific reasoning and involves some dependence on empirical knowledge. This dependence has indeed been asserted, in the case of geometry, at least from the time of Kant; and the relativists of to-day, when they do not maintain that physics is geometry, declare geometry to be physics. But arithmetic, etc., which have more readily passed for pure and *a priori* sciences, are deemed by some philosophers to be in the same case as geometry. The question has been discussed with acuteness and great learning by M. Émile Meyerson in his latest book, *Du Cheminement de la pensée*, and I would here reproduce his thought on one or two points.

An equation is not an identity. That 7 and 5 are 7 and 5 is an identity; that 7 and 5 are 12 is an assertion of partial identity, equality or equivalence, between the diverse. And this latter assertion, or the simplest instance of mathematical reasoning, involves a constructive operation of the mind, concealed by the word 'and'. The diversity between the terms here concerned is so slight as to seem not to exist; we judge that no mind can refuse consent to the conclusion expressed by the sign=; and we say 7 and 5 *are* 12. Even here identification involves elimination of the negligible. Numbers are originally got by abstraction from things, and it is because this class of abstractions provides so wide a field of application, in the procedure of identification, that arithmetic appears to be, from the first onwards, an *a priori* science. In the higher stages of mathematical reasoning ideal experimenting is the form often taken by the constructive mental activity. This is discernible in Galileo's thought and in some of Euclid's methods of proof, and it plainly involves memory of actual or empirical observations and experiments, or perhaps a medley of memories. Immediacy, as I have remarked in other connexions, is often reducible to familiarity, or to the capacity to guess results after experience or practice. The new in mathematical thought indeed often comes by way of induction; and the mathematician's mind, in making an advance or a discovery, has proceeded sometimes according to intensional logic, or with the notion of coherence of attributes in an essence, while formal demonstration by extensional logic is an afterthought—an instance of being wise after the event. Again, justification by results, or by successful application, rather than by strict deduction from clearly defined concepts, commends instruments such as the infinitesimal calculus, and enabled Poincaré to regard d'Alembert's scruples with regard to it as almost incredible.

These disconnected observations may serve to shew

that there is some reason to believe that mathematics and the physical sciences are not so disparate, in respect of the processes of reasoning by which they have been developed, as has been wont to be thought. Elimination of the negligible can be carried further in mathematics, and, what is more important, what is neglected is much more obviously negligible. It is by this fact that M. Meyerson accounts for the greater rigour of mathematical demonstration, despite its use of constructive reasoning essentially similar to that employed in the physical sciences. In them, as we know, neglect, which is one side of identification, is tentative: it can lead astray, or prove wrong, so that what was accounted negligible sometimes needs to be reinstated. In mathematics, where abstraction goes to its uttermost length, and leaves us with empty concepts, this risk is generally *nil*: the neglect of properties is no cause of error. Having relegated the concrete to a distance at its first step, mathematics has no need to reintroduce it. Hence its easy and certain procedure, and its prestige in respect of rationality.

Lecture VI

THE RELATION OF THEOLOGY TO OTHER DEPARTMENTS OF KNOWLEDGE

In a previous lecture I dealt with the relation of dogmatic theology to one department of knowledge—historiography. But the theology of revealed religion has been asserted to have another basis for its truth-claim besides historiographical data, viz. the experience of Christians in any generation. In this last lecture of my course I propose to set forth, with some comment, the views which have been held by typical schools of theology, natural as well as revealed, as to the foundations of their science and its basis in religious experience, and as to whether theology is independent of other departments of knowledge.

If I may use the word 'modernity', in connexion with theology, to mean freedom of thought from external authority, rejection of the traditional infallibilities, and adoption of verifiability rather than apostolicity or œcumenicalness as the criterion of truth, theology may be said, without much qualification, to have entered on modernity with eighteenth-century deism. The free lances who were named deists, but who would be more accurately called rational theists, seem to be the first to put into practice, in theology as distinct from institutions, etc., that independence to which the Reformation had asserted man's right. From their time onwards some few theologians have regarded theology as an outgrowth from our knowledge concerning the world and man. But the majority have represented it to be an isolated science, independent of all or nearly all other departments of thought. Within the 'modern' period, as I am now using that word, the endeavour to divorce theology from the secular sciences and philosophy arose in the romantic reaction against the dry-light 'Illumination' of the eighteenth century, which repudiated all alleged 'inner light' save that of reason, and narrowly conceived of reason as but *a priori* intuition and ratiocination. The tendency was doubtless also caused by the criticism which the classic proofs of theism and the system of ecclesiastical doctrine received from Hume and Kant. It gathered strength in the nineteenth century by including a protest against the intrusion of Hegelian intellectualism into theology.

This tendency to isolate theology is a good instance of history repeating itself. Indeed, in essential respects it is so closely analogous to a movement of thought which had worked itself out in an earlier period that description of it can be aided by instituting a parallel.

If scholasticism, in its essence as distinct from its exposition in the schools, may be said to have begun with Erigena, it entered on its course of developement with a calm confidence in the unity of all truth, the identity of theology with philosophy, and of faith with reason. "True philosophy is true religion, and true religion is true philosophy", said the first great medieval theologian. Nearly three centuries later this intellectualism could still be embraced by Abelard. But before the identity of theology with philosophy thus received its second enunciation it had already appeared less obvious to Anselm. Anselm taught that reason could justify the dogmas of the Church, though they stood in no need of its patronage, and so could lead faith on to knowledge; but he recognised that churchly faith and philosophic knowledge are distinct, and that a logic is needed to bridge the chasm between their respective spheres. Another century elapses, and the dialectic which Anselm had deemed sufficient is pronounced to be inadequate. It breaks down when called upon to supply rational demonstration or deduction of the more mysterious doctrines of the Church. In the age of the summists the hope of theoretically proving such dogmas was abandoned. Aquinas acknowledged that there are truths above reason; and he would make room for them by narrowing the domain of knowledge. To try to prove by natural reason the Trinity of Persons or the Incarnation was to detract from the rights of faith. Thus was recognised a twofold truth, or the distinction between natural theology and revealed theology. Thenceforward the decline of intellectualism was rapid. Duns Scotus maintained the supremacy of the will over the intellect, upheld the practical nature and the pragmatic verifiability of religion, and insisted on its independence of reason and philosophy. Soon the rational demonstrability of the being of God, firmly believed in by Aquinas, was surrendered by William of

Ockham. He, as the radical empiricist of his day, resigned all knowledge transcending experience to the sphere of faith. Finally some representatives of scholasticism in its decline, not to speak of forerunners, seem seriously to have accepted the notion of the 'double truth'. Not only are reason and faith, philosophy and theology, distinct: what is false in the one may be true in the other. Thus one of the conclusions reached by an age-long movement of thought was the exact opposite of the belief as to the rationality of faith, with which the movement began.

Returning now to recent history, one may speak of Hegel as the modern Erigena. Before his day reason had again shewn itself unwilling to accept Bacon's eirenicon and to "render unto faith the things that are faith's". Kant, for instance, had treated the historical facts and the doctrines of Christianity as symbols and parables, the real significance of which only appears when they are rationalised, or regarded as illustrative of ethical principles. Reason had retorted to faith faith's own authoritative word to it; "Whom ye ignorantly worship, Him declare I unto you". But never, since Abelard, had the hope of speculative thought mounted so high as during the developement of the idealism in which the rationalistic side of Kant's philosophy attained its consummation. Hegel's system assigned, in its dialectically outlined evolution of the Absolute, a place to religion such that its whole meaning and essence were theoretically determined. Religion, Hegel taught, sheltered no mystery which thought could not disclose:

religion is absolute knowledge. Thus the first systematic 'philosophy of religion' given to the world reproduced the confident intellectualism of the first Christian rationalist, if with a different construing of Christian dogma. And we may observe how closely the recoil of theology from Hegel's position parallels the medieval movement away from the corresponding standpoint of Erigena.

The first stage in the gradual descent from Hegel's rationalism, as distinct from Schleiermacher's abrupt plunge, was represented by Lotze. This philosopher denied the adequacy of thought to comprehend the whole of reality without remainder. He believed metaphysics to be capable of supplying a proof of theism, or of the existence of an intelligent world-ground, but regarded theoretical knowledge as incapable of making any further advance towards religion. He held that considerations as to worth were necessary to effect the transition from philosophy to theology. But just as the common-sense position, if I may so call it, of Aquinas afforded no abiding-place for the restless age which followed, so the corresponding position of the almost Christian Lotze came to be abandoned by theologians whose thought was disturbed by the intellectual upheavals of the nineteenth century. Quondam disciples of Lotze were led on to more and more complete denials of the rights of reason within the domain of religious belief. The long dominant school of Ritschl, which at first owed much to Lotze, would credit metaphysics, or theoretical knowledge, with no capacity to find out God, and would regard religious faith as exclusively rooted in the practical side of our nature. Ritschl sought to derive theology solely from what he called judgements of worth, and to eject from dogma all that could not be so derived: for instance, the deity of Christ was said to involve no such metaphysical assertion as that He is of one substance with the Father, or is the pre-existent Logos, but simply that He has "the value of God". The verification of a theological doctrine, again, was said to be no matter of dialectics or metaphysical reasoning, but to consist in the capacity of the doctrine to explain religious experiences and to promote the spiritual life. In this component of the Ritschlian theology we may perhaps see the source of the pragmatist notion of truth, which invaded other fields than that of theology, and which became a cardinal feature of the type of 'modernism' taught by Loisy. This, however, is to anticipate. Ritschl's divergence from Lotze is essentially a repetition of the Scotist revulsion from the position of Aquinas. The collapse of medieval rationalism in Ockham finds its modern analogue in the tendency, fostered by Schleiermacher, William James, and other writers, to restrict the scope of the philosophy of religion to a psychology of religious experience and an inductive investigation of facts supplied by that science and by a comparative study of religions. Theology would thereby seem to be cancelled from inquiries into ultimate truth, and to lapse into a department of anthropology. Nevertheless, many theologians who follow psychological rather than rationalistic methods believe those methods capable of leading to supersensible knowledge, and also that, in detaching the sources of theological doctrine from the subject-matter of other departments of knowledge, they have placed theology out of the range of all hostile artillery. One of the leading members of the Ritschlian school went so far as to assert that whether natural philosophy issues in materialism is indifferent to a Christian; and some 'modernists', in the technical sense of the name, proclaim that the truth of the doctrines and ideas of Christianity is unassailable, though many of the alleged facts on which traditional Christianity professes to be based are indemonstrable, and may even be fictious. Such professions encourage one to pursue the parallel which I have been drawing to its bitter end. That faith can afford to scorn alike the support and the conceivable hostility of knowledge seems perilously similar to the paradox of decadent scholasticism. And unless the new teaching can make good its fundamental claims, when confronted with what we have learned concerning the scope and validity of such belief as we are wont to call knowledge, it would seem that it must inevitably be an unintentional re-assertion of 'the double truth'.

The movement of thought of which I have given a brief history has, in the main, been an endeavour to isolate, in one way or another, theology from other fields of knowledge and thought. The profane sciences, as we have seen, possess one broad characteristic in common. Their original matter, or the objective nucleus of their primary data, is the sensorily perceptual.

Our percepts and their simpler relations are the sources of our ideas and universals, at least in the sense of being the occasions of our obtaining them, though, once thus derived, ideas can be related without further recourse to the sensory, so as to yield the pure sciences. It is by percepts in the first instance, and then by other objects derived from or by means of them, that feeling and valuing are evoked, so that appreciation as well as cognition, and consequently the sciences of valuation, such as ethics, ultimately presuppose sensory data. Again, sense-givenness is the sole original certificate of actuality; and it is the perceptual that gives us our first, our most direct, if phenomenal, touch with the real that underlies the phenomenal. If all knowledge is thus ultimately conditioned, on its objective side, by sensation, it follows that theology must be an outgrowth from ordinary knowledge of the world and man. This, however, is just what those who regard theology as an isolated science are concerned to deny. And one of the grounds on which the denial is based is that theology sets out from objective data of a different genus from that of the sensory and the sense-derived. These experience-data are said to be as original, i.e. as ultimate or irreducible, as sense-impressions, to be as immediately apprehended as they are, and to afford another but equally first-hand touch with the real—i.e. the ontal—world as that which our sciences of the perceptual enjoy. Thus theology has claimed access to an additional realm of actuality, or a spiritual environment, though the actuality of this realm has not the vouch of sense-givenness. This realm is credited with the capacity to excite a unique kind of emotional response and of valuation, whereby religious experience *sui generis* is constituted; and theology is described by many of its exponents as the systematisation of this peculiar type of experience.

If this claim could be substantiated, the profane sciences would indeed not be, even potentially, the whole of knowledge. Theology would supply a further contribution, and one of which it should behove philosophy, or an organisation of our knowledges, to take account. In respect of its peculiar data theology would be independent of the profane sciences and, so long as it confined itself to its own sphere, should be immune from either scientific or historical criticism. It is natural that those who regard theology as thus a science of unique experiences should glory in its isolation, seeing therein a guarantee of its invulnerability.

The forthcomingness of experience-data of another order and kind than that on which our profane knowledge is based is, of course, a possibility not precluded by the fact that profane knowledge makes no use of them and succeeds in its own tasks by proceeding as if there were no such data. But whether this possibility is also an actuality, whether analysis of religious experience, as we find it organised at the level of common sense, results in the discrimination of data at once immediate, ultimate, and unique, and whether religious experience in general is inexplicable and indescribable without invocation of such alleged data, these are questions which demand critical examination from the analytical and genetic science of knowledge-processes, which is as fundamental for theology as it is for philosophy. They raise several critical issues which I will proceed to indicate.

There is a general consensus of writers on religious experience as to one point: viz. that the peculiarity of the affective or responsive side of religious experience is determined by the peculiarity of its objective side. Those who hold that religious experience is characterised by a single kind of sentiment or a single type of valuation agree that the uniqueness of that sentiment is due to the uniqueness of the object or objects which elicit it. Various things, or objects of profane knowledge, suggestive of mysteriousness and power, have been religiously reverenced by mankind, or have evoked valuation distinguishable from other types such as the aesthetic and the moral. Stones and wells, trees, animals, the human dead, the heroic personality, and the moral law have been carriers or vehicles of the sacred, the numinous, or the divine. It is this invisible sacred presence, as distinct from its visible vehicle, which is generally said to be the religious object proper. It pervades its carriers, and is what makes them secondarily sacred. Just as tears are not exhaustively described or explained when they are said to be composed of water and salts, so the visible vehicle of the sacred is not exhaustively described or explained when science has told us all she knows about it. The vehicle is mysterious and evokes numinous or religious valuation; but this, it is said, is so because it is indwelt by the numen. The numen, or religious object proper, is further declared to be an existent with which there is acquaintance, in the technical sense of that word, or of which there is apprehension as immediate as is sensory acquaintance with a visual object. Such, in generalised form, is the claim of theology, of the non-mystical type, to be in possession of distinctive data, the apprehension of which bespeaks a normal intuitional functioning of the human mind essentially different from any involved in ordinary knowledge of the world and man. Theology based on abnormal and distinctively mystical experience differs only in asserting acquaintance with the numinous as not lodged in perceptual things, but it does not now call for separate examination.

There are two grounds at least on which this claim may be said to be disputable. Firstly, the kinds of acquaintance-knowledge which we undoubtedly possess consist in apprehension of concrete qualities such as colour; and it is not easy to conceive of acquaintance with actual 'terms', as distinct from subsistent relations, which is not correlative with presentation of some specific quality. The numinous object, however, has no specific quality such as that which constitutes a colour or a sound; it is characterised only by its agency, which is not an object of immediate apprehension, in causing or evoking a specific kind of valuation or subjective attitude. One shrinks from saying that acquaintance with the non-qualitative is inconceivable or self-contradictory only because some philosophers profess to know their own egos by acquaintance, and pure egos are agents devoid of any apprehensible *quale* analogous, e.g. to blueness. But if this theory as to self-consciousness be inconsistent with facts, or superfluous because a more natural one is forthcoming, it will furnish no reason for accepting the notion of immediate apprehension, akin to sensation, of the insensible. It would seem then that if there are numinous beings, one or more, they are comparable with Kant's things *per se*, Leibniz's monads, or the physicist's electrons, rather than with directly presented objects such as coloured surfaces. And if this objection to the view that the numinous is directly apprehended be considered to be indecisive, another presents itself which seems fatal to it. For it should follow, since acquaintance precludes illusion, that all the powers and deities, of all the mythologies and religions from the crudest Nature-worship to monotheism, are alike real; or else that the one Being that has manifested itself to numinous experience has selectively indwelt, for the purpose of self-revelation, this, that, and the other kind of physical or human object, the derived sacredness of which was only obvious to a particular community. On the former of these alternatives being adopted, theology would be involved in self-refutation; and if one adopts the latter of them one is confronted with the fact that progress from the lower forms of religion to ethical monotheism has largely consisted in successive refusals to regard as habitations or manifestations of God the objects that were accounted sacred in more primitive times. Thus it would seem that the numinous object, constitutive of religious experience throughout its many stages of refinement, cannot be a quasi-perceptual datum, of the same order of underivedness as the sensory. Its vagueness and lack of quality, in virtue of which it can figure in all kinds of mythology and theology, and in all kinds of religious experience mystical or normal, bespeak its identity with the generic image or with the concept reached by abstraction and idealisation, rather than its affinity with objects of first-hand apprehension or acquaintance.

The second crucial point involved in the claim of theology to unique data is the alleged immediacy with which these data are said to be received. Mysticism and the intuitionist theology founded by Schleiermacher alike rely on the immediacy of religious experiences, while some theologians see in it a basis for dogmatic theology securing its independence of historiography. Consequently much hangs on the sense in which immediacy can be claimed, and the question needs further investigation than it has as yet received.

Truly immediate apprehensions are distinguishable from many which persons unversed in analytic and genetic psychology have called immediate: e.g. conclusions arrived at by inference, interpretations reached by constructive imagination or conceptual synthesis of which the experient is unconscious, and traditional opinions to which familiarity has imparted the semblance of self-evidence. These latter experiences are immediate for the subject who has them only in the sense that he is unaware of their mediateness; and that is irrelevant to the issue before us. As I have had occasion to observe in previous connexions, there are two standpoints from which immediacy can be asserted, or two notions of immediacy. Nowhere is it more important to recognise the distinction between them than in discussions of religious and mystical cognition, and nowhere, I am afraid, has it been more generally overlooked. The immediacy alleged of his experience by the religious experient is really discerned from the one standpoint and affirmed from the other. And the important question is not whether mediatedness happens to be undiscerned by a given experient at a given moment, but whether it is found to be absent from an actual situation when that is scientifically explored and analysed. An assertion that the latter is the case, made without the requisite inquiry, is worthless; and when the inquiry is made, the assertion seems superfluous and untenable. For the transcendent object of religious experience turns out, as I have previously submitted, not to be a quasi-impressional datum, uniquely apprehended with genuine immediacy, but rather a derived and mediated image or conception which is interpretatively read into perceptual or ideal objects, as the case may be. It is *thought* to be there, or is suppositionally assigned as the *unapprehended* cause of mental states, such as emotions and sentiments, uplifitings, and so forth, which are immediately apprehensible in introspection. What is called God-consciousness must in that case be, like self-consciousness, an indirectly and reflectively acquired, and an interpretative, kind of knowledge or belief.

Schleiermacher has been regarded as a theologian of epoch-making importance on account of his express denial of this conclusion. He taught that we have an intuition of the manifold of finite things as a connected whole, or a unity, existing as such only through the infinite and eternal One manifested in that unity. He asserted this intuition to be immediate, and regarded its immediacy as securing the distinctness of piety from science and morality. But the intuition in question plainly presupposes a system of abstract ideas, indeed science and philosophy, and is mediated by such knowledge. Mankind certainly got the notion of the transfinite only after much study of the finite, and the notion of an absolute One only after discerning relations within the many. It seems to me difficult to imagine a more extravagant abuse of the word 'immediate' than this of Schleiermacher's. Nevertheless it is but an extreme case of an error that is by no means the monopoly of theologians. Men are apt to think that acts of conceiving which they now perform with acquired facility are as immediate as if they were innate; and I have previously criticised a typical instance of this tendency, selected from the almost common-sense philosophy of Professor Sidgwick.

Returning to theology, I would now indicate the third point involved in the assertion that theology is an isolated science in virtue of its unique data. Every experience, whether cognitive or affective, is one in which a subject is necessarily confronted with an object. Religious experience is undoubtedly *rapport* with an object. But from this psychological truth the theologians who would base their science on what are called "the inwardly verifiable facts of the soul's experience" have leaped to the metaphysical conclusion that religious experience is self-evidenced touch with a real spiritual environment. It is then overlooked that objectivity, in the psychological sense borne by the term in their indisputable premiss, is not coextensive with actuality or reality which is asserted in their conclusion. The psychologically objective includes the imaginal and the ideal, as well as the actual: the centaur and the Euclidean line are objective, in that they are not subjective states or activities. Moreover imaginal and ideal objects, when they are believed to be actual, can evoke feelings and sentiments as profound, intense, inspiring, and practically fruitful, as those excited by perceptual or actual things: and imaginary persons count for quite as much as real ones in the lives of most people. These feelings, etc. cannot, therefore, without further ado, be said to evidence the reality, or the non-ideality, of the objects which excite them. And here, by the way, is indicated the fatal flaw in all pragmatist inferences from the spiritual fruitfulness of an experience to its implication of real objects, or from the efficacy of a doctrine to its truth. If theology based on religious experience would account itself a department of knowledge of the actual and the ontal, the ability to assert that that experience has its objective side is not enough.

There remains the further question, whether there is a real or actual counterpart to such objects as religious experience indubitably involves. And that question cannot be answered by religious experience itself. For religious experience has begged it in the very act of constituting itself religious, or of unique kind. Religious experience, it is generally admitted, contains no ultimate analytica that do not enter into other kinds of experience, save the alleged quasi-impressional but insensible data which I have already submitted reasons for believing to be derived, conceptual, and read in, rather than genuinely immediate or read off. Consequently there remains no way of accounting for the uniqueness of religious experience save by attributing it to the introduced interpretative idea of God, or of the numen. This idea permeates the other data or analytica, and it alone bestows upon them the capacity to evoke emotional response of a peculiar kind. Previously to the acquisition and the causal or interpretative use of this derived notion, experiences such as were destined to become religious could not be religious: they could only be regarded as natural, not as supernatural—whether aesthetic, moral, or of other types. It is this cognitive element, originally anthropic interpretativeness, which mediates the religiousness of religious experience: just as substance and cause, which are likewise interpretative ideas and not directly apprehended impressions or relations, mediate science, as distinct from sporadic sensations, etc. Thus, when it is said that theology is derived from religious experience it is nevertheless true that, primarily and fundamentally, religious experience presupposes the theological concept of the divine or the numinous, and owes its uniqueness to saturation with that concept. That which it is sought to derive from the data of religious experience has all the time been read into such apprehensions as are genuinely immediate, constituting them religious data; and these data are not pure data but data *plus* an interpretative factor.

I have laboured this point with some repetition because, whether my view be right or wrong, the question involved is one of fundamental importance, and the issue needs to be made clear. It has been obscured in consequence of several confusions which I have sought to expose: those, viz., of genuine with spurious immediacy, of psychological objectivity with reality or actuality, of pure data with interpreted data, and of the mental state of efficacious certitude with logical or scientific certainty. On one or two of these false identifications, to which the experience-school of theology is addicted, a few more words may be said.

Theological pragmatism is summed up in such a statement as that a religion under the influence of which a genuine spiritual life has flourished cannot be simply false. Now a highly developed religion is a complex thing and usually comprises ethics as well as theological or metaphysical doctrine. And inasmuch as the phrase 'spiritual life' includes ethical goodness, a religion may promote spiritual life in virtue of its purely ethical teaching. But when we speak of a religion as true, we generally mean true as to its dogma. Its ethics may be lofty, and may be moulded by worthy doctrinal ideas, and yet those ideas may conceivably have no counterpart in actuality. It must not be ignored that a person's convincedness as to the truth of a dogma is a sufficient condition of its efficacy in ministering to his spiritual life, and that the truth of the *credendum* is then a superfluous condition. The *credendum* may conceivably be ungrounded or even false, while convincedness as to its truth is fruitful. At any rate we must not confound a *credendum* with belief in it, nor read the causal agency involved in the promotion of spiritual life into the object believed in, without asking whether that agency may not be located in the experient's believing. Until the latter explanation is proved impossible or inadequate we have no right to assume the former to be dictated by the facts of experience.

The mistaking of what are called the data of religious experience for pure, or wholly read off, data, is perhaps more common in academic circles than the fallacy of theological pragmatism. When, for instance, the Christian asserts his immediate 'sense' of the indwelling Christ, he is asserting more than he can immediately apprehend in introspection. What alone he so apprehends consists of consolations, upliftings, bracings of will, joy, peace, etc. On these genuinely immediate experiences or pure data he imposes a causal interpretation which he has only come to possess through instruction in Christian dogma. And he would not be able to extract that dogmatic content out of his present experiences had it not been first interpretatively read into them. This dogmatic content may of course be true, but it is not matter of direct experience in the sense that joy and peace are. To take another illustration: remission of sin has been said to be a fact such as renders idle any attempt to dismiss the theological doctrines of sin and grace. The phrase 'remission of sin' is then plainly intended to imply a supersensible Remitter of sin, and perhaps some doctrine of atonement, over and above describing the actual experience of feeling oneself no longer overwhelmed with the debilitating sense of guilt and shame. So here again an agency and an agent which are not directly apprehended are read into pure fact-data, such as emotional states, and regarded as if in the same sense 'given' to immediate apprehension as those data are given. It is not any natural science, such as biology, having no need of the notions of sin and grace, which confronts dogmatic theology presuming to be based on living or genuine experience, but rather the analytical psychology of cognition.

If theology is not derivable from religious experience because religious experience already presupposes a theological and interpretative factor derived elsewhere, it follows that it is by a more circuitous path than the short cut of alleged immediacy, and by trespass on property other than that of religion as confined to alleged unique data, that theology must arrive, if it can arrive, at beliefs such as other sciences would account reasonable. I will discuss this possibility presently.

Meanwhile, I would briefly deal with the other main ground on which it has been claimed that theology is an isolated science. Theology has been declared to be independent of some of the other fields of knowledge, such as natural science and metaphysics, if not of all existential propositions other than its own, in that its fundamental doctrines can be extracted out of value-judgements alone. It is not easy to ascertain exactly what some of the upholders of this view mean. For sometimes it appears to be only certain traditional systems of metaphysics, and not metaphysics as such, from which it is sought to isolate theology. And sometimes propositions asserting existence are referred to as if they were identical with abstract and necessary truths. But, what is more directly relevant here, the phrase 'value-judgement' seems to be a misnomer for a predication of existence as well as of worth, or of an actuality possessing value. Then a so-called value-judgement figures as a premiss surreptitiously retaining the existential or metaphysical element which this kind of theology professes to have strained out of its premisses, but which it would fain re-establish as a conclusion from value-judgements strictly so called. But if it is difficult to ascertain what the apparently confused statements of the Ritschlian school mean, it is plain enough what they should mean if their intention is the isolation of theology. They should mean that theological doctrine is implied in purely ethical propositions and is independent of any world-view suggested by the generalisations of physical science, or speculatively obtained by philosophy in its endeavour to co-ordinate our various departments of knowledge. And, when thus clearly stated, the intended conclusion seems clearly to be false. Even if ethical principles and ideals were absolute, or independent of all cognition and human interests, we should need to be assured that the universe respects our ideals and aspirations before we could argue from their present forthcomingness to their perpetuation and fulfilment. The value of goodness is one thing and the permanent existence of the good is another. So far as ethics can know, the moral life of man and his advance in culture may be but a transient episode in a blind and meaningless cosmic evolution. There is thus no direct implication, in man's possession of moral dignity, ethical principles, categorical imperatives, etc., of the existence of a Deity and a suprasensible world. Unless theism, or something like it, be first established, we have no reason to believe that ethical ideals, such as the highest good, are destined to be realised. And if, apart from theism, hope to that effect is groundless, we cannot use the forthcomingness of the ideals as a premiss from which to deduce God and immortality. The principle of the conservation of value, in which some would see the essence of religion and theology, is not a postulate of the world's rationality in the sense in which science finds the world rational, but only of rationality of the teleological kind; and whether the world is rational in the latter sense can only be ascertained by study of the world. When it is said that the *facts* of our moral life imply God, grace, eternity of life, etc., or that the reality of our moral pilgrimage involves the reality of the goal, facts would seem to be identified with ideas, their actual preconditions to be confounded with the logical preconditions of a value judgement, and the order of being with the conditions and processes of idealisation. In other words, it is only by reviving a form of the ontological argument that realisation of the highest good can be extracted out of the idea of the highest good and its oughtness to be. Kant was shrewd enough to see this. He did not profess to provide a moral proof of theism, but merely argued that the idea of God is a postulate for the practical reason as well as a regulative idea for the theoretical reason. He taught that *if* the moral order is to stand, and the highest good or what ought to be is to be, then God must exist. It is only more audacious venturers than Kant who have sought to deduce the permanent standing of the moral order from the deliverances of the moral consciousness; and their attempts seem always to involve the confusion of the realm of the valid with the realm of the existent. If theism can be substantiated by appeal to facts of the same order as those which must first have suggested the theistic idea or its cruder forerunners, it will solve the final problem of philosophy—the relation of what is to what ought to be; but it seems to me that from ethics alone there is no more a direct approach to theology than there is from the genuine immediacies, in that mixture of genuine and spurious immediacies of which religious experience consists. And, again, the latter type of experience will receive its most adequate explanation if theism can be antecedently or independently established. That the peace which passes understanding is possessed by some people is fact; but that it is given by God to those who believe in Him is a certitude which cannot be experimentally or experientially proved to be a certainty, any more than it admits of disproof by an unbelieving experimentalist.

From the grounds on which it has been sought to secure for theology the isolated position of an independent science I may now pass to those on which it may be rather conceived as the final link in a continuous chain of interpretative belief. This is not the occasion for elaborating a theistic argument, but it may be suggested that the world of which we have scientific knowledge may be found to admit of reasonable explanation only in terms of theism. By reasonable explanation I mean something different from a logically and positively established world-view; for no such product, whether theistic or atheistic, is possible. I mean something different also from acquiescence in the forthcomingness of an astounding complication of happy coincidences in our world without asking how or why. The cosmos may strongly suggest that it is ordered for the realisation of moral and other values. To that extent natural science would itself supply the reasonable guarantee which ethics or religious experience alone cannot provide, and would redeem religious belief or hope from groundlessness. The very sciences from which certain schools would cut theology loose may thus afford the only indisputable facts, by building on which theology can become a body of reasonable beliefs for the guidance of life. And certainly one may say that Butler's analogy between natural and revealed religion might be supplemented to-day by an analogy between natural theology and science. For inductive science has its interpretative explanation-principles, as we have previously observed, and its faith-elements with which the faith of natural theology is, in essence, continuous. The hard line that once was wont to be drawn between scientific knowledge and theistic belief has been considerably softened since the analytical and genetic study of scientific knowing entered on its modern phase. Science, as we have seen, is not positive or apodeictic; not necessary, unconditional or universal; not adequate or exhaustive. Its verification is pragmatic, not logical. And perhaps no advance in knowledge has been of more moment in connexion with the relation of science, as a whole, to natural theology, than the recent explication of the metaphysical assumptions, concerning substance and causation, involved in the identification of problematic inductions with certain truth. The probability which is the guide of science turns out to be ultimately the same in logical and psychological nature as that which is the guide of life and of reasonable prudence. The faith involved in theism such as is based on cumulative teleological considerations is essentially the same as that belief in the world's rationality which is presupposed by the logic and method of science, and theistic belief is but a continuation, by extrapolation, or through points representing further observations, of the curve of 'knowledge' which natural science has constructed. In short, science and theism spring from a common root. Not a single item of genuinely scientific knowledge would need to be disputed if the real world were proved to be purely spiritual and primarily a realm of ends; and science has no call to warn off other kinds of assimilative or interpretative explanation such as shift the emphasis to aspects of Nature which science is not concerned to regard—e.g. from physical magnitudes to spiritual values, or from calculabilities to spiritual significance. The only broad differences between science and theology are in respect of their data and the degrees in which verification is possible within their spheres. Subjective activities and beliefs are involved in the data of all common knowledge, such as science; but religious experience seems to be conditioned, both as to its existence and its distinctive nature, by further antecedent belief over and above such as is indispensable for knowledge of the physical. Yet this element of over-belief is not scientifically unreasonable. It receives its justification when we pass from religious experience, and dogma founded on it, to the natural theology which the other kind of theology presupposes, and find natural theology, or theism, to be but a continuation of science as it is constituted before its main line of procedure from the historical takes on the direction towards the abstractly mathematical. As to verification of their respective postulations, again, there is some difference between theology and science. In science, verification consists in appeal to the external control of percepts, whereas the results by which religious postulation is pragmatically justified are concerned with valuation and edification rather than with existential facts. And the religious postulates are not so inevitable, *prima facie*, as those of physical science: they involve a further, though not an essentially different, venture of faith. But, again, when we pass from first appearances to remoter considerations, the difference seems to become largely one of degree rather than of kind.

Such I take to be the relation between theology and other departments of knowledge as it presents itself to the natural theologian. Theology explicates what the other departments and sciences suggest; and they supply it with a basis, in facts and generalisations, for a faith such as is but a further stage, in that venture to believe where we cannot rigidly prove, which we have found to be inevitable in all that we are wont to call knowledge of actuality. It is a case, the natural theologian may say, of "united we stand, divided we fall". Natural theology, apart from the sciences, is baseless: natural science, stopping short of a theistic culmination, has the appearance of an arbitrarily arrested growth. Theology is not an isolated nor an isolable science; it is an outgrowth of our knowledge of the world and man. Revealed theology presupposes natural theology, and natural theology has no data other than those which experience supplies to science.

Without attempting a recapitulation of this course of lectures I may bring it to an end by indicating a few of the more general conclusions which have been reached.

The analytic and genetic study of knowing, as an actual process, reveals that knowing consists in interpreting the primary or first-hand manifestations of the real or ontal world to our subjecthood, with its specifically human faculties and capacities. Between the world to be known and knowledge of it stands human nature; and between an individual man and science stands the race, or sociality. We can perhaps eliminate anthropomorphism, but we can never transcend the anthropic. We can, by various supposings, ventures, and expediencies, pass from individual or private experience to common or over-individual knowledge; but by no manner of means can we transcend the common and arrive at over-social, absolute, or dehumanised touch with reality. Our departments of knowledge differ most significantly in respect of the number and kind of interpretative categories and explanation-principles that are indispensable for their respective reductions of the chaos of brute and historical facts to system; and thus they may be arranged in an order. But there is no linear advance from some one fundamental concept to others increasing in complexity. The concepts and categories involved in knowing are *ad hoc*, and are often mutually implicative. And those required by the more historical sciences, as indispensable to their procedure, are invoked with as much right as are the fewer and more formal categories by the purer sciences, once it is recognised that all knowledge of the actual world is pre-eminently interpretative. The principle of parsimony cannot be applied to the use of categories without loss of the significance of the particular order of facts which a particular science studies, nor without loss of the significance of that science for the whole of knowledge.

But if it is thus necessary to divide in order to conquer, it is also necessary to recognise such continuities and interdependencies between the sciences as are forthcoming, if we would consolidate our conquests. Our knowledge, as a whole, is comparable to an organism with its members rather than to a house with its walled-off rooms; our departments of knowledge are not compartments. We cannot separate the sciences save in a somewhat superficial manner. The study of the objective side of common experience, abstracted from the subjective side and in isolation from knowledge as to the faculties of the human soul, may lead to science; but it will not lead to natural philosophy or to metaphysics or to a reasonable view as to the world and man. It is idle, we have seen, to raise metaphysical questions before studying what knowledge actually is, or to proceed as if the intelligibility of the world to man did not depend as much on anthropic interpretativeness as on the structure and nature of the ultimate reality with which man is in touch. When we reflect upon our knowledge as a whole we discover that science presupposes psychology and theory of knowledge though it ignores them in pursuing its own business. It also presupposes the historical, which history systematises in terms of continuity and developement, and which science itself seeks to 'understand' in another sense, involving less of comprehension as understanding approximates to computation. Indeed history, in the broader sense of the name, and the science of knowledge are the first sciences in the systematic order of presupposition which a philosophy of the sciences establishes. Metaphysics can only be approached through them. The pure sciences, on the other hand, supply no premisses from which truth valid of the actual world can be extracted. They are concerned with the ideal, with postulated 'entities', but not with the ontal which reveals itself in producing the phenomenal, nor with the actual save in so far as numerical and other formal relations subsist between things.

The relations between the sciences which I have selected for discussion imply relations between the processes and actualities with which those sciences or departments of thought are respectively concerned. For instance, the body, which physiology describes, mediates to the soul, which is the subject-matter of psychology, its knowledge of the world, with which physics deals, and its primary categories, which are the concern of epistemology. And it is this inter-connectness of things which urges the human knower to seek an explanation, reasonable rather than rational, of their manifold inter-adaptations. The sciences lead intellectual curiosity on to philosophy. And when philosophy finds its explanation in the supposition that the world and man constitute an organic whole, whose ground is God and whose *raison d'être* is realisation of the good, it passes into natural theology. Such, I take it, is the relation in which theology stands to the other departments of knowledge.